TMDLs in the Urban Environment

Case Studies

SPONSORED BY
Task Committee for Urban TMDLS

Environmental and Water Resources Institute (EWRI)
of the American Society of Civil Engineers

EDITED BY
Stuart M. Stein, P.E.

Published by the American Society of Civil Engineers

Cataloging-in-Publication Data on file with the Library of Congress.

American Society of Civil Engineers
1801 Alexander Bell Drive
Reston, Virginia, 20191-4400

www.pubs.asce.org

Acknowledgments

The Task Committee for Urban TMDLs formed in 2001 with the purpose of collecting representative case studies of urban TMDLs in order to facilitate future TMDL development. The Task Committee's leadership consisted of the following individuals:

Stuart M. Stein, Chair
David D. Dee, Jr., Co-Chair
Gordon B. England, Secretary
Leslie Shoemaker
Andrew Parker

In pulling together the case studies included herein, Mr. Stein worked closely with Leslie Shoemaker and Andrew Parker. The Task Committee would like to recognize the contributions of the authors of each of the case studies, without whom this publication would not be possible:

Don Waye
Judy Buchino
Brian Watson
Drew Ackerman
Terrence Fleming
Ken Schiff
Stephen Carter
Tom Henry
Rui Zou
Leslie Shoemaker
H. S. Greening
A. J. Janicki

Additionally, the Task Committee would like to thank Robert Traver. Robert was elected chair of the Urban Water Infrastructure Committee while our Task Committee was actively producing this document and he provided invaluable assistance in shepherding the publication through proper EWRI channels.

Contents

Introduction

This document presents several detailed urban Total Maximum Daily Load (TMDL) case studies. These TMDLs have been accepted by the appropriate regulatory authority (the State environmental permitting agency for those States with permitting authority) and so they are examples of acceptable modeling, acceptable pollutant allocations, and, in some cases, acceptable implementation plans. In presenting these case studies, we are not endorsing their technical adequacy; rather, we present the case studies to inform stakeholders of permitting agency expectations for urban TMDLs, which will compel actions taken in developing these TMDLs.

The National Pollution Discharge Elimination System (NPDES) is the dominant regulatory driver for urban runoff water quality management. It is critical that stakeholders understand the potential link between the two programs since an accepted TMDL can obligate mitigation activities undertaken in compliance with NPDES permits.

Case Study Summaries

This document presents four case studies of the efforts to develop and implement TMDLs. The case studies represent a range of geographical locations, watershed sizes and characterstics, land use characteristics, pollutants, TMDL development methods, and TMDL implementation. They are briefly summarized below and in Table 1.

Table 1. Summary of TMDL Case Studies

Stream	State	Landuse	Pollutant(s)	Main Source	Suggested Implementation
Four Mile Run	VA	Urban	Fecal coliform	Wildlife (waterfowl and raccoons)	Repair sanitary sewer lines, employ street sweeping, control nuisance wildlife, etc.
LA River	CA	Mixed	Nutrients (nitrogen)	Point sources, WWTPs	Reduce discharges at WWTPs
Tampa Bay	FL	Mixed	Nutrients (nitrogen)	Point Sources, WWTPs	Reduce discharges at WWTPs
Wissahickon Creek	PA	Urban	Nutrients and siltation	Point Sources, WWTPs (nutrients), streambank erosion (siltation)	Reduce discharges at WWTPs (nutrients), reduce runoff volume (siltation)

WWTP = Wastewater Treatment Plant

Four Mile Run, VA - Bacteria

The 19.7 mi^2 Four Mile Run watershed drains Arlington and Fairfax counties, and the cities of Falls Church and Alexandria, VA, and empties into the tidal portion of the

Potomac River. The watershed is primarily urban and over 40% impervious, leading to a variety of flow and quality degradation issues. Of these, the pollutant of greatest concern is bacteria, which has led to the development of a TMDL for fecal coliform bacteria. Genetic fingerprinting of E. coli was used to characterize the source species of fecal contamination and help quantify loading. Hydrologic Simulation Program – Fortran (HSPF) was used to model the watershed loadings and concentrations of fecal coliform contamination and identify opportunity for appropriate best management practices and policy actions to meet the TMDL. Management strategies include reducing sanitary sewer overflows, more street sweeping, and better wildlife management.

Los Angeles River, CA – Nutrients (Nitrogen)

The 819 mi^2 Los Angeles (LA) River watershed drains a large area of Southern California, is fed by several major tributaries and empties into San Pedro Bay. The land use in the watershed is highly varied, ranging from undeveloped forest to high density residential, commercial and industrial. Many of the reaches of the 55-mile LA River suffer from nutrient impairment, and associated effects such as low dissolved oxygen and eutrophication. Nitrogen (nitrate, nitrite, ammonia, etc) sources are distributed throughout the watershed, but three major primary point sources (Wastewater reclamation plants) account for as much as 80% of the loading, especially during low flow periods. A TMDL was developed for both the nutrient (Nitrogen) concentrations and associated effects. The Environmental Fluid Dynamics Code (EFDC) linked with the Water Quality Analysis Simulation Program (WASP) was used to simulate the LA River and nutrient loadings. The modeling helped to establish waste load allocations for the three point sources to help meet the TMDL in flow periods.

Tampa Bay Estuary, FL – Nutrients (Nitrogen)

The Tampa Bay Estuary is a large estuary (1000 km^2, or 386 mi^2) which drains a mixed land use watershed of over 5700 km^2 (2200 mi^2). High loadings of Nitrogen have caused significant algal growth and eutrophication of Tampa Bay, leading to a considerable reduction in extent and quality of seagrasses and other submerged aquatic vegetation. The Tampa Bay Estuary Program (a multi-agency consortium) developed a TMDL with Total Nitrogen as the pollutant of concern. The sources of nitrogen are diverse, including point sources, non-point sources, atmospheric deposition, and agricultural runoff, among others. A public private consortium has been developed to identify actions to control watershed nitrogen loadings and meet seagrass habitat quantity and quality goals.

Wissahickon Creek, PA – Nutrients and Sediment

Wissahickon Creek drains 64 mi^2 of Southeastern Pennsylvania in a formerly agricultural area that is now mostly urbanized. The urbanization has led to significant changes in nutrient loadings and flow conditions. The stream suffers from low dissolved oxygen, eutrophication, and heavy siltation. As a result, two separate

TMDLs have been developed: one for nutrients and one for sediment. Nutrients enter the stream from nonpoint sources and point sources, notably WWTPs. Through WASP/EUTRO modeling, EPA determined the TMDL for nutrients should include ammonia, nitrate and nitrite, orthophosphate, and CBOD as pollutants. The primary reduction strategy for these pollutants is to reduce permitted discharges for the 23 NPDES permitted dischargers, though only five were expected to be significantly affected. The sediment loadings were attributed primarily to nonpoint sources and MS4 permit dischargers instead of NPDES permit holders. Since MS4 permit holders are not regulated based on sediment, a range of regulatory and non-regulatory is being implemented to control stormwater runoff and reduce peak flows to reduce silt deposition from overland flow and streambank erosion.

Overview of Current Total Maximum Daily Load-TMDL-Program and Regulations

The following is a summary of the TMDL program provided on the Environmental Protection Agency's (EPA) website (www.epa.gov/owow/tmdl).

The Need - The Quality of Our Nation's Waters

Over 40% of our assessed waters still do not meet the water quality standards states, territories, and authorized tribes have set for them. This amounts to over 20,000 individual river segments, lakes, and estuaries. These impaired waters include approximately 300,000 miles of rivers and shorelines and approximately 5 million acres of lakes-polluted mostly by sediments, excess nutrients, and harmful microorganisms. An overwhelming majority of the population-218 million-live within 10 miles of the impaired waters.

Section 303(d) of the Clean Water Act

Under section 303(d) of the 1972 Clean Water Act, states, territories, and authorized tribes are required to develop lists of impaired waters. These impaired waters do not meet water quality standards that states, territories, and authorized tribes have set for them, even after point sources of pollution have installed the minimum required levels of pollution control technology. The law requires that these jurisdictions establish priority rankings for waters on the lists and develop TMDLs for these waters.

What is a TMDL?

A TMDL specifies the maximum amount of a pollutant that a waterbody can receive and still meet water quality standards, and allocates pollutant loadings among point and nonpoint pollutant sources. By law, EPA must approve or disapprove lists and TMDLs established by states, territories, and authorized tribes. If a state, territory, or authorized tribe submission is inadequate, EPA must establish the list or the TMDL. EPA issued regulations in 1985 and 1992 that implement section 303(d) of the Clean Water Act - the TMDL provisions.

Litigation

While TMDLs have been required by the Clean Water Act since 1972, until recently states, territories, authorized tribes, and EPA have not developed many. Several years ago citizen organizations began bringing legal actions against EPA seeking the listing of waters and development of TMDLs. To date, there have been about 40 legal actions in 38 states. EPA is under court order or consent decrees in many states to ensure that TMDLs are established, either by the state or by EPA.

Overview of the 1992 TMDL Regulations-Under Which the Current Program Operates

- **Scope of Lists of Impaired Waters**

  o   States, territories, and authorized tribes must list waters that are both impaired and threatened by pollutants.

  o   The list is composed of waters that need a TMDL.

  o   At the state's, territory's, or authorized tribe's discretion, the waterbody may remain on the list after EPA approves the TMDL, or until water quality standards are attained.

  o   States, territories, and authorized tribes are to submit their list of waters on April 1 in every even-numbered year, except in 2000. In March 2000, EPA issued a rule removing the requirement for the 2000 list - though some states are choosing to submit such lists on their own initiative.

- **Methodology Used to Develop Lists**

  o   States, territories, and authorized tribes must consider "all existing and readily available water quality-related information" when developing their lists.

  o   Monitored and evaluated data may be used.

  o   The methodology must be submitted to EPA at the same time as the list is submitted.

  o   At EPA's request, the states, territories, or authorized tribes must provide "good cause" for not including and removing a water from the list.

- **Components of a TMDL**

  o   A TMDL is the sum of allocated loads of pollutants set at a level necessary to implement the applicable water quality standards, including -

       ▪   Wasteload allocations from point sources.

       ▪   Load allocations from nonpoint sources and natural background conditions.

       ▪   A TMDL must contain a margin of safety and a consideration of seasonal variations.

- **Priorities/Schedules for TMDL Development**

  o States, territories, and authorized tribes must establish a priority ranking of the listed waterbodies taking into account the severity of pollution and uses to be made of the water, for example, fishing, swimming, and drinking water.

  o The list must identify for each waterbody the pollutant that is causing the impairment.

  o States, territories, and authorized tribes must identify waters targeted for TMDL development within the next two years.

- **Public Review/Participation**

  o Calculations to establish TMDLs are subject to public review as defined in the state's continuing planning process.

- **EPA Actions on Lists and TMDLs**

  o EPA has 30 days in which to approve or disapprove a state's, territory's, or authorized tribe's list and the TMDLs.

  o If EPA disapproves either the state's, territory's, or authorized tribe's list or an individual TMDL, EPA has 30 days to establish the list or the TMDL. EPA must seek public comment on the list or TMDL it establishes.

- **1997 Interpretative Guidance for the TMDL Program**

  o EPA issued guidance in August, 1997, to respond to some of the issues raised as the program developed. The guidance includes a number of recommendations intended to achieve a more nationally consistent approach for developing and implementing TMDLs to attain water quality standards. These recommendations include:

    ▪ States, territories, and authorized tribes should <u>develop schedules</u> for establishing TMDLs expeditiously, generally within 8-13 years of being listed. EPA Regions should have a specific written agreement with each state, territory or authorized tribe in the Region about these schedules. Factors to be considered in developing the schedule could include:

      ▪ Number of impaired segments.

      ▪ Length of river miles, lakes, or other waterbodies for which TMDLs are needed.

      ▪ Proximity of listed waters to each other within a watershed.

      ▪ Number and relative complexity of the TMDLs;

      ▪ Number and similarities or differences among the source categories;

      ▪ Availability of monitoring data or models; and

      ▪ Relative significance of the environmental harm or threat.

- States, territories, and authorized tribes should describe a plan for implementing load allocations for waters impaired solely or primarily by nonpoint sources, including -

    - Reasonable assurances that load allocations will be achieved, using incentive-based, non-regulatory or regulatory approaches.

    - Public participation process.

    - Recognition of other watershed management processes and programs, such as local source water protection and urban storm water management programs, as well as the state's section 303(e) continuing planning process.

The National Pollution Discharge Elimination System (NPDES)

Urban stormwater management actions are, for the most part, dictated by NPDES. The following is a summary of information on the NPDES stormwater program provided on the Environmental Protection Agency's (EPA) website (www.epa.gov).

Stormwater discharges are generated by runoff from land and impervious areas such as paved streets, parking lots, and building rooftops during rainfall and snow events that often contain pollutants in quantities that could adversely affect water quality. Most stormwater discharges are considered point sources and require coverage by an NPDES permit. The primary method to control stormwater discharges is through the use of best management practices.

Under the NPDES storm water program, operators of large, medium and regulated small municipal separate storm sewer systems (MS4s) require authorization to discharge pollutants under an NPDES permit.

Medium and large MS4 operators are required to submit comprehensive permit applications and are issued individual permits. Regulated small MS4 operators have the option of choosing to be covered by an individual permit, a general permit, or a modification of an existing Phase I MS4's individual permit.

Operators of regulated small MS4s must have permit coverage no later than March 10, 2003. Under the Small MS4 Stormwater Program, operators of regulated small MS4s are required to:

- Apply for NPDES permit coverage.

- Develop a stormwater management program which includes the six minimum control measures.

- Implement the stormwater management program using appropriate stormwater management controls, or best management practices (BMPs).

- Develop measurable goals for the program.

- Evaluate the effectiveness of the program.

Listed below are the six minimum control measures that operators of regulated small MS4s must incorporate into stormwater management programs. These measures are expected to result in significant reductions of pollutants discharged into receiving waterbodies.

- Public Education and Outreach.

- Public Participation/Involvement.

- Illicit Discharge Detection and Elimination.

- Construction Site Runoff Control.

- Post-Construction Runoff Control.

- Pollution Prevention/Good Housekeeping.

Regulatory Implications for Urban Areas

As previously stated, the NPDES requirements are, to a large extent, driving local urban runoff water quality management decisions. Importantly, this permitting system is not independent of the TMDL program. According to a memorandum dated November 22, 2002 from the Director of the US EPA Office of Wetlands, Oceans, and Watersheds to the Water Division Directors of US EPA's 10 regions, NPDES-regulated storm water discharges must be addressed by the wasteload allocation component of a TMDL and NPDES permit conditions must be consistent with the assumptions and requirements of available wasteload allocations. In other words, TMDLs are to be treated as part of the applicable NPDES permit. This implies that failure to meet TMDL conditions could possibly result in financial penalties (fines) under NPDES. These fines can be significant–upwards of $50,000 per day per violation depending upon several factors.

While TMDLs have traditionally been "paper" exercises, specifying allocations without enforcing action to meet these allocations, the link between TMDLs and NPDES stormwater permits greatly strengthens enforcement authority. While NPDES Phase II is a relatively new program and permitting agencies are displaying some flexibility in order to assist permit holders in properly implementing their permits, enforcement action will eventually be taken on those permit holders not meeting permit conditions. Potential future NPDES enforcement action will eventually result in more proactive implementation of urban TMDLs.

Fecal Coliform TMDL (Total Maximum Daily Load) Development for Four Mile Run, Virginia

Donald Waye[1] and Judy Buchino[2]

[1]USEPA Headquarters, Ariel Rios Building, 1200 Pennsylvania Avenue, NW
Mail Code: 4503T, Washington, DC 20460, 202-566-1170, waye.don@epa.gov

[2]Northern Virginia Regional Commission, 7535 Little River Tnpk., Suite 100,
Annandale, Virginia 22003, 703-583-3828, www.novaregion.org

Abstract

A TMDL determines the maximum amount of pollutant that a water body is capable of receiving while continuing to meet the existing water quality standards. The Virginia Department of Environmental Quality (VADEQ) listed the Four Mile Run watershed in Arlington and Fairfax County, Virginia on the Commonwealth's 1998 303(d) TMDL Priority List of Impaired Waters, thus requiring a TMDL to be set for fecal coliform.

There are several potential sources for fecal coliform in the Four Mile Run watershed. The watershed is urban and is approximately 40% impervious, so all but agricultural sources could contribute to the fecal load. Bacterial source tracking and computer modeling tools were used to identify the main sources of fecal coliform in the watershed. The findings point to wildlife, mainly waterfowl and raccoons, as the primary fecal coliform source in the watershed.

Various scenarios were developed with modeling tools to determine what load reductions are necessary to meet Virginia water quality standards. DEQ intends for this TMDL to be implemented through best management practices (BMPs) in the watershed. Implementation will occur in stages

TABLE OF CONTENTS

LIST OF TABLES

LIST OF FIGURES

1. Introduction

1.1 Background

Section 303(d) of the Federal Clean Water Act and the United States Environmental Protection Agency's (USEPA) Water Quality Planning and Management Regulations (40 CFR Part 130), requires states to identify water bodies that are in violation of the water quality standards for any given pollutant. Under this rule, states are also required to develop a Total Maximum Daily Load (TMDL) for the impaired water body. A TMDL determines the maximum amount of pollutant that a water body is capable of receiving while continuing to meet the existing water quality standards. TMDLs provide the framework that allows states to establish water quality controls to reduce sources of pollution with the ultimate goal of water quality restoration and the maintenance of water resources.

The Virginia Department of Environmental Quality (VADEQ) listed the Four Mile Run watershed on the Commonwealth's 1998 303(d) TMDL Priority List of Impaired Waters (VADEQ, 1998). Four Mile Run is a direct tributary of the Potomac River and is located in Virginia River Segment VAN-A12R, which is a portion of the Shenandoah-Potomac River Basin that drains into the Chesapeake Bay.

1.1.1 Study Area Description

Four Mile Run is an urban stream that spans most of Arlington County and parts of three other localities: Fairfax County, the City of Alexandria, and the City of Falls Church. The stream flows from west to east, with a slight southerly tilt. This TMDL addresses a fecal coliform bacteria impairment identified by VADEQ that begins at the headwaters of Four Mile Run just over nine miles upstream of its confluence with the Potomac River to the tidal/non-tidal boundary approximately 1.5 miles upstream from the Potomac. Figure 1 shows the location of the Four Mile Run watershed. While the entire watershed is 19.7 square miles, the nontidal portion of the watershed covered in this TMDL is 17.0 square miles.

There is no agricultural runoff in the watershed, which is home to 183,000 people, or just over 9,000 per square mile (NVRC staff analysis of 2000 U.S. Census data). The dominant land use in the watershed is medium to high density residential housing.

Not surprisingly, Four Mile Run has a higher daytime population during the workweek than its 183,000 permanent residents. The watershed is approximately 40% impervious. Aside from a crowded human populous, there is a large pet population in the watershed. In addition to these two sources, the 1998-2001 study of bacteria sources in Four Mile Run by the Northern Virginia Regional Commission (NVRC) and Virginia Tech illustrate the influence of waterfowl (Canada Geese and mallards, in particular) and raccoons as sources of E. coli. Figure 2 provides a summary pie chart of this study's findings.

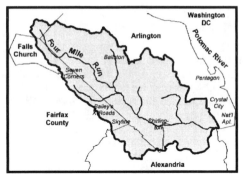

Figure 1. The Four Mile Run Watershed in Northern Virginia

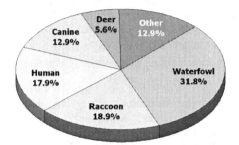

Figure 2. Isolate matches from NVRC's BST investigation in Four Mile Run with Virginia Tech, 1998 - 2001

In recent years, five groups have performed fecal coliform monitoring of Four Mile Run—VADEQ, NVRC, the Fairfax County Health Department, the Arlington County Parks Division, and the Arlington Chapter of the League of Women Voters. All have found elevated levels of fecal coliform bacteria in the Four Mile Run watershed. Since 1990, over 700 fecal coliform samples have been taken from Four Mile Run and its tributaries. Nearly half of these samples have been determined to be over the 1,000 most probable number (MPN) Virginia water quality standard for fecal coliform bacteria.

Importantly, there is little manufacturing industry to generate point source discharges. While there are two regulated point source discharges in the watershed, one is a small concrete batch plant with a pH discharge regulation only and the other is Arlington's modern sewage treatment plant (STP), which provides tertiary treatment and easily complies with its 200 colony forming units (cfu) per 100 milliliters (mL) permit limit for fecal coliform bacteria (NVRC analysis of Arlington ST daily discharge monitoring records, 1998 – 2001). This plant discharges in the tidal portion of Four Mile Run near the Potomac River, and is thus outside the study area of this TMDL.

In the summers of 1999, 2000, and 2001, NVRC performed optical brightener monitoring (OBM) on each of the 297 outfalls in the watershed, many of which were monitored more than once. OBM is a technique that has been used in rural watersheds and the caves of the Ozarks to successfully trace human sewage to its source. The results revealed two isolated problems of moderate severity, which were corrected quickly, and eight outfalls with possible low-level contamination of human sewage for which investigations are ongoing.

While conducting monitoring for its municipal separate storm sewer system (MS4) permit in 1998, Arlington County staff discovered an illegal cross-connection from a condominium complex in Fairfax County that discharged to a stream in Arlington, and a repair was quickly made. Fieldwork for OBM and MS4 monitoring has also revealed intermittent problems typical of heavily urbanized watersheds, such as improper dumping of wastes. While OBM monitoring is limited by its ability to detect only human sewage that contains laundry waste, its findings, along with visual and "sniff" observations at every outfall in the watershed reveal a stream with little obvious direct human sewage component.

1.2   Impaired Water Quality Status

VADEQ determined that Four Mile Run exceeded one of the existing instream fecal coliform water quality standards and identified the source of impairment as being urban nonpoint source runoff. Fecal coliform bacteria are the primary resident bacteria in the feces of all warm-blooded animals. Although it is not usually pathogenic, fecal coliform bacteria is commonly used as an indicator for potential health risks resulting from pathogenic organisms that are also known to reside in feces. The Four Mile Run watershed has been given a TMDL status of "medium priority" resulting from the Virginia Water Quality Assessment for 1996 and a high NPS ranking in VADEQ's 1998 305(b) report to Congress and EPA.

1.3   Goal and Objectives

The goal of the Four Mile Run TMDL is to allocate the sources of fecal coliform contamination and to incorporate practices that will reduce fecal coliform loads and allow Four Mile Run to meet Virginia state water quality standards. The following objectives must be completed in order to achieve this goal:

Objective 1—Assess the water quality and identify the potential sources of fecal coliform

Objective 2—Quantify current fecal coliform loads and estimate the magnitude of each source

Objective 3—Model and predict the current fecal coliform loads being deposited from each source

Objective 4—Develop allocation scenarios that will reduce fecal coliform loads

Objective 5—Determine the most feasible reduction plan that can realistically be implemented and incorporate it into the TMDL.

2. Watershed Characterization

2.1   Climate

The Four Mile Run watershed straddles the Mid-Atlantic piedmont and coastal plain physiographic provinces approximately 50 miles east of the Blue Ridge Mountains, and 35 miles west of the Chesapeake Bay. Watershed elevations range from sea level to 425 feet above mean sea level. Four Mile Run is a tributary of the Potomac River, and enters the river on its western shore at the southern end of Ronald Reagan National Airport (formerly Washington National Airport). The primary sources for information presented throughout this section are documents and records from the National Weather Service (NWS).

Climate data for this area have been kept continuously since November 1870. Official observations have been recorded since June 1941 at Reagan National Airport. This airport is at the center of the urban heat island associated with the greater Washington, D.C. area. Consequently, low temperatures recorded at the airport are approximately 10 to 15 degrees higher than the surrounding suburban areas (NWS, 2002). The recorded high temperatures are not as greatly affected by the urban heat island effect, so there is less variation in high temperature readings between urban and suburban locations.

Winters are usually mild, with an average temperature in the mid 20's (°F). Spring and fall are generally mild climates, with very pleasant weather. Summers can be hot and humid, with temperatures averaging about 80°F. The average date of the last freeze in spring is April 1, and the average date for the first freeze in the fall is November 10.

Precipitation is generally evenly distributed throughout the year, with an annual rainfall of 39 inches per year. Snowfalls average 18 inches per year, with perhaps only one or two major snowfalls in a season. It is unusual to have a snowstorm of 10 inches or more within any one particular day. However, there have been rare occurrences of 25-inch snowstorms.

Frost (1998) analyzed the historical rainfall record around Washington, D.C. over a 96-year period and identified four distinct types of precipitation events: trace, convective, frontal and cyclonic. An analysis of each rainfall event from 1972 through 1976 revealed that frontal systems accounted for 37% of the total number of storms and 39% of the total volume of precipitation over the five-year period. Trace events were the second-most common type of precipitation, accounting for 28% of the events, but only 3% of the volume. 25% of the events were generated by warm weather convective cell atmospheric disturbances, which accounted for 24% of the volume. Finally, cyclonic systems produced only 10% of the storm, but 34% of the volume.

2.2   Land Use

Land use is a predominant determining factor for source of fecal coliform deposition. For example, wildlife is more common in open space and parkland than highway corridors and high-density development. Likewise, pet populations are associated with residential lands more so than commercial or industrial areas.

Land use information was obtained from NVRC's own Northern Virginia regional land use GIS layer with a multi-jurisdictional 15-key land use classification. A sixteenth land use category was culled from this GIS layer by parsing major highways from the "Public Open Space" category they shared with open parkland. Other minor cleaning of this layer was performed to ensure the final accuracy of this important model input. It should be noted that two land uses in this regional GIS layer are absent from the watershed—open water and rural residential/agricultural. Thus, the model uses 14 land uses. The determination and distribution of watershed imperviousness is derived from this supplied land use information. Thus, attention to the quality of this land use information is a large reason the hydrology calibration, described later, has an exceptionally good fit.

The nontidal portion of the Four Mile Run watershed is 10,874 acres, or 17.0 square miles. Table 2-1 shows the acreage of each existing land use in the impaired portion of the watershed and the average estimated impervious land use. Land use acreage is also broken down for each of the three segments delineated for the Four Mile Run TMDL computer model (presented in Chapter 4). Using Table 1 yields an overall imperviousness for the impaired portion of the watershed of 41.5%. This value is consistent with other estimates from watershed localities and NVRC's Four Mile Run SWMM Model, which place the watershed within the 35 to 45 percent impervious range.

Table 1. Land Use Classification by Model Segments in Acres

Land Use	Impervious	Seg1	Seg2	Seg3	Total
Open Space/Parks	2%	390	180	40	610
Highway	90%	213	126	130	469
Medium to High Density Mixed Use	65%	241	80	96	417
Medium to High Density Industrial	80%	24	110	20	154
Public/Conservation/Golf	8%	148	102	309	559
High Density Residential	75%	20	179	101	300
Medium Density Residential	40%	2,692	755	804	4,251
Medium to High Density Residential	50%	392	930	414	1,736

Table 1. Land Use Classification by Model Segments in Acres (con't)					
Land Use	Impervious	Seg1	Seg2	Seg3	Total
Medium to High Density Commercial	70%	86	69	100	255
Low to Medium Density Residential	20%	767	243	33	1,043
Low Density Commercial	40%	260	274	7	541
Low Density Industrial	65%	9	46	5	60
Low Density Mixed Use	30%	12	189	0	201
Federal	50%	0	100	178	278
Total		5,254	3,383	2,237	10,874

2.3 Water Quality Data

Four Mile Run water quality data used for the development of this TMDL was compiled from the following sources:

- Virginia Department of Environmental Quality (VADEQ)
- Arlington County Department of Parks, Recreation, and Community Resources (DPRCR)
- Northern Virginia Regional Commission (NVRC).

The VADEQ data has been collected at least quarterly and at most semi-monthly at a single station in the nontidal portion of Four Mile Run since 1991. Prior to this, some sampling by VADEQ was performed during the 1970s, but this sampling was discontinued by 1980. VADEQ's identifier for this station is 1AFOU004.22, and it is located along the Four Mile Run mainstem directly under the Columbia Pike (Virginia Route 244) bridge. Throughout this report, this station is referred to as Four Mile Run at Columbia Pike.

Data collected by the Arlington County DPRCR supports its annual put-and-take trout stocking program in Four Mile Run. County park naturalists collect fecal coliform bacteria data, along with dissolved oxygen and pH, to gauge stream conditions leading up to opening day of trout season, which is usually in late March. As a result, a variable number of samples are collected from early February to mid-March most years at four locations along Arlington's greenway park system that straddles the middle section of Four Mile Run's mainstem. Unfortunately, however, no data was collected by DPRCR during calendar year 2000, and only one value was obtained for calendar year 2001. One of the DPRCR stations, designated as FMR3, is located approximately 800 feet upstream Four Mile Run from Columbia Pike. As there are no tributaries or other significant drainage between FMR3 and Columbia Pike, and the reach is reasonably uniform along this section, data collected at this location was deemed appropriate to include with the other observed data collected at Four Mile Run and Columbia Pike. All data collected at Columbia Pike and FMR3 during the period simulated by the TMDL model (January 1, 1999 through May 31, 2001) was used for calibration and verification.

Five fecal coliform values were collected by NVRC and Virginia Tech at Columbia Pike and Four Mile Run during the period simulated by the TMDL model described in Chapter 4. This data was collected to support the NVRC/Virginia Tech BST study.

These datasets can be characterized by the percent of the violations of Virginia's instantaneous standard of 1,000 cfu/100 mL. Table 2 shows the frequency of violation of the instantaneous fecal coliform standard by source and location from 1991 through the most recently available data.

2.3.1   Seasonal Analysis

Seasonal variation for instream fecal coliform concentration was performed for Four Mile Run. The seasonal cutoffs used in this analysis were the actual calendar dates for each season, and were not rounded by month.

Table 2. Fecal Coliform Standard Violation Frequency in the Four Mile Run Watershed

Source	Location(s)	Years	# of Observations	Frequency of Violations for Instantaneous Standard*
VADEQ	Four Mile Run at Columbia Pike	1991 - 2001	41	27%
Arlington County Parks	4 sites along Four Mile Run mainstem from Bon Air Park to Barcroft Park	1998 - 2002	63	14%**
NVRC	29 sites throughout nontidal portion of watershed, including tributary streams	1998 - 2000	42	33%
All Sources	Combined	1991 - 2002	146	23%

- * 1,000 counts (most probable number) per 100 mL of stream water
- ** Arlington limits data collection to late winter (February to mid-March) in association with its annual trout stocking program. See Table 3 for seasonal distributions.

**Table 3. Fecal Coliform Standard Violation Frequency by
Data Source and Season**

	Frequency of Violations for Instantaneous Standard*					
	VADEQ		NVRC		VADEQ + NVRC + Arlington	
	%	# of obs.	%	# of obs.	%	# of obs.
Winter	20%	10	20%	10	16%	83
Spring	33%	12	60%	15	46%	27
Summer	25%	8	25%	8	25%	16
Fall	27%	11	11%	9	20%	20
Overall	27%	41	33%	42	23%	146

* 1,000 counts/100 mL

Results show that the mean fecal coliform concentrations for the samples collected by the VADEQ are above the instantaneous standard for three seasons: winter, summer, and fall, with the highest mean values occurring during the fall season. The high winter mean fecal coliform concentration of 1,353 for the VADEQ data is attributable to a single reading of 7,800 MPN on February 17, 1999. Excluding this value results in a drop of the winter mean to 636.

3. Source Assessment

3.1 Nonpoint Sources

3.1.1 Bacteria Source Tracking (Genetic Fingerprinting)

The development of this TMDL greatly benefited from a significant genetic fingerprinting investigation on the DNA of E. coli in the Four Mile Run watershed performed by Dr. George Simmons of Virginia Tech's Biology Department from 1998 through 2000. A technical paper on this study was published in a peer-reviewed book titled *Advances in Water Monitoring Research*, released earlier this year (Simmons, 2001). Field data for this source tracking study was collected on five separate trips to the watershed at 31 different locations and across all four seasons. Some locations were visited on multiple occasions, and the number of DNA matches varied from site to site based on a number of different factors.

Genetic fecal typing, or BST, represents one line of evidence; long-term observations by trained naturalists working in the watershed represent another. Following the release of the BST results, NVRC performed in-depth interviews with five top naturalists working in and near the Four Mile Run watershed. The purpose of these interviews was to ascertain the degree of overlap between bacteria sources suggested by the source tracking study and what the naturalists believed the sources should be. The interviews revealed near 100% agreement among the naturalists on which sources should be found in the watershed and their relative numbers and habitats, as well as which species were likely to be absent from the watershed, or in some cases, seasonally absent.

While information from these interviews revealed a large degree of overlap with the DNA evidence, some disparities emerged. For example, certain waterfowl species (e.g., least tern and black back gull) implicated by DNA evidence were believed by all five naturalists to be absent from the watershed year-round. Where these two lines of evidence contradicted each other, DNA matches were reclassified as "disputed" for the purposes of developing this TMDL.

Fortunately, not only were the disputed cases limited to a few problem species, the overall DNA results track closely with a similar BST study (using RNA fingerprinting) in the Accotink Creek watershed performed in 2000. The centroids of these watersheds are approximately 10 miles apart, and their land uses are roughly similar.

The percentages of the resulting classifications after the disputed matches were removed were used as a starting point and guide for modeling source contributions. As a practical matter, the percentages for each modeled segment could not be used directly in the model. For example, the number of isolate matches is so low for Segment 3 (lower nontidal Four Mile Run) that no matches were found for humans, raccoons or canines, despite their populations being in roughly the same proportions as found in Segments 1 and 2. There is considerably closer agreement in the proportion of waterfowl and raccoons between Segments 1 and 2, and the higher sample sizes of these segments make their percentages less suspect.

While the human and canine percentages show much more variability across Segments 1 and 2, the genetic tools applied in this study has difficulty distinguishing between bacteria strains from these two host species. However, because of the persistent nature of human matches found at one particular storm drain outfall at the upper end of Doctors Run in Model Segment 2, coupled with consistently high bacteria counts obtained at that location by NVRC in this study and its subsequent investigation, NVRC suspects this to be a hotspot for human bacteria sources. In short, percentages of sources derived from the DNA source tracking investigation served as a guide for model loadings, along with information from the naturalists and NVRC's own long track record of analysis from fieldwork and census and pet records for the watershed.

3.2   Point Sources

There are no permitted or known point source discharges of bacteria in the watershed. Two of the four localities that share the watershed - Arlington and Fairfax counties - have municipal separate storm sewer system (MS4) permits. The other two localities - the cities of Alexandria and Falls Church - are expected to receive MS4 permits within the next few years. The permits blur the lines that have traditionally distinguished point and nonpoint sources of pollution. While the MS4 permits are regulated similarly to point source discharges, water quality discharging from the MS4s is nearly exclusively dictated by nonpoint source runoff (along with an unknown, but presumed small, amount of illicit connections). In the Four Mile Run watershed, the MS4s intercept groundwater flow during baseflow periods, and are dominated by runoff during and immediately after rainfall. This baseflow is controlled by pervious surface processes such as infiltration, while the storm flow is dominated by runoff from impervious surfaces. Optical Brightener Monitoring (OBM) conducted by NVRC staff from 1999 to 2001 at every outfall in the watershed lends evidence that storm sewer outfalls are largely free from illicit connections (NVRC, 2000; and various in-house OBM project documents, 1999-2001). This evidence is supported by Arlington County's MS4 monitoring results over the past three years on file with VADEQ.

4. Modeling Approach for Four Mile Run Total Maximum Daily Load

The most critical component of Total Maximum Daily Load (TMDL) development is to establish the relationship between the source loadings and the in-stream water quality. This relationship is essential for the evaluation and identification of management options that will achieve the desired source load reductions. Modeling the relationship between loads and water quality can be achieved through different techniques ranging from simple mass balance models to more sophisticated dynamic and fully integrated watershed scale modeling. However, when the fate and transport of a pollutant depends upon the changing responses to runoff flow and source loadings, it is important to use a model that simulates the loadings from various non-point sources and characterizes the resulting stream water quality for the different runoff and stream flows that may occur in the watershed.

This section describes the steps to select a model and to develop the information needed to apply the model to hydrologic and water quality simulations of Four Mile Run. It details the modeling tools used, the existing physical and hydrologic data, the hydrology approach used for the calibration, the development of direct and indirect source loadings used in the water quality model, and the approach used for the water quality calibration of the model.

4.1 Model Description

The model selected for Four Mile Run is HSPF—Hydrologic Simulation Program – Fortran. HSPF is a set of computer programs that simulate the hydrology of the watershed, nutrient and sediment nonpoint sources loads, and the transport of these loads in rivers and reservoirs. HSPF partitions the watershed into three smaller sub-watersheds (upper, middle and lower Four Mile Run). Data on land uses and nonpoint sources are entered into the model for each sub-watershed. The primary interface for this application of HSPF is WinHSPF and full advantage of EPA's BASINS modeling environment was taken in the development of key components of this model. However, the Four Mile Run HSPF model also benefited by moving beyond the somewhat limited data inputs and calibration options available through the interfaces offered by BASINS and WinHSPF.

In its production run configuration, the Four Mile Run HSPF model generates daily nonpoint source edge-of-stream pollutant loads for each land use and instream concentrations at each sub-watershed outlet. Each sub-watershed contains information generated by a specific component or submodel. Results from the three submodels (hydrologic submodel, non-point source submodel, and river submodel) combine to estimate the changes in load estimates to Four Mile Run. The hydrologic submodel uses rainfall and other meteorological data to calculate runoff and subsurface flow for all the watershed land uses. The runoff and subsurface flows, generated by the hydrologic sub-model, ultimately drive the nonpoint source sub-model. The nonpoint source sub-model (PERLND and IMPLND) simulates multiple pathway transport of pollutant loads from the land to the edge of the stream. The river sub-model (RCHRES) then routes flow and associated pollutant loads from the land through the stream network to the outlet of the watershed.

4.2 Model Sub-watershed Discretization and Land Use

The Four Mile Run watershed was divided into three sub-watersheds that are identified as Segment 1 - upper Four Mile Run; Segment 2 - middle Four Mile Run; and Segment 3 - lower nontidal Four Mile Run. They are often referred to in tables by the shorthand "Seg1," "Seg2," and "Seg3." Figure 3 illustrates this sub-watershed division and sampling station locations. The sampling station location between Seg1 and Seg2 on this map is the VADEQ monitoring site at Columbia Pike (1AFOU004.22). The sampling station between Seg2 and Seg3 is the USGS stream gauge at the Shirlington Road bridge crossing of Four Mile Run. The Shirlington station was used to calibrate the hydrologic response of the model, and the Columbia Pike station was used for bacteria calibration. The dot at the eastern edge of Seg3 is the tidal/nontidal downstream boundary of the TMDL model.

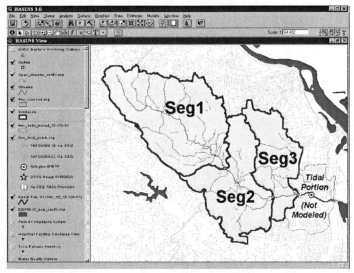

Figure 3. Subbasin Divisions for the Four Mile Run TMDL Model Segmentation

The locations of available flow and bacteria data to calibrate the model were the primary considerations for determining sub-watershed model boundaries. The sole acceptable stream gauge data set is from the U.S. Geological Survey (USGS) flow gauge on Four Mile Run at the Shirlington Road Bridge. High resolution flow data (at 5- to 15-minute intervals) was collected from October 1998 through the present, and is even available in near-real time online at <waterdata.usgs.gov/va/nwis/uv?01652500>. The only two long-term fecal coliform monitoring stations in the nontidal portion of the watershed are the one operated by the Virginia DEQ at Four Mile Run and Columbia Pike and one operated by the Fairfax County Health Department in the headwater portion of upper Long Branch— a tributary to Four Mile Run. The tributary site, located near the Fairfax/Arlington county line, was considered by NVRC to have too small of a drainage area to warrant its own HSPF model segment, and was therefore not useful for model calibration. The outlet for HSPF Model Segment 1 is the DEQ monitoring site at Columbia Pike and the outlet for Model Segment 2 is the USGS stream flow gauge in Shirlington.

High-resolution, ground-truthed land use information exists in standard digital GIS formats, and was generated by a previous NVRC project. The automated land use and model segmentation capabilities of BASINS were used to automatically extract information from the land use layer and add them to the HSPF model for each sub-watershed segment in correct model input format. The segment-specific land use information was presented in Table 2-1.

4.3   Selection of Model Simulation Period

Because neither hourly nor daily flow data exists prior to October 1998, and because of the start-up period required by HSPF, the model calibration period was from January 1, 1999 through May 31, 2001. Continuous hourly time series inputs for precipitation, air temperature, dewpoint, potential evapotranspiration, and wind speed were added to the model input stream from July 1, 1998 to May 31, 2001. Most of these inputs exist for both Reagan National Airport at the lower end of the watershed and for a Fairfax County Health Department weather station in Seven Corners at the upper end of the watershed.

Because insufficient data existed to test the model calibration parameter values against a separate verification period, the 29-month calibration period was subdivided into two periods for the purposes of providing a mini verification exercise. That is, while the final calibration parameter values were derived based on the period of January 1, 1999 through May 31, 2001, separate calibration statistics were also tracked for the periods January 1 – December 31, 1999 and January 1, 2000 – May 31, 2001. Calibration results for these two periods were very similar. Additionally, calibration statistics were tracked for seasonality—again with no evident seasonal bias in the final calibration results. Results of this calibration exercise are presented later in this chapter.

4.3.1   Availability of Precipitation Data

Precipitation is a particularly critical model input and serves as the primary driver for simulating stream flow and bacteria densities. Not all precipitation stations operated rain gauges continuously during the period of simulation. Thus, only the gauges at Seven Corners and Reagan National Airport were used as model inputs. Small gaps in the Seven Corners dataset were filled with hourly precipitation records from a station approximately one mile northwest at Sisler's Stone (a store) in Falls Church.

4.4   Hydrology Modeling Approach

This section describes the approach used for the hydrology model calibration in Four Mile Run. Simulating the long-term hydrologic response requires extensive information on the physical, meteorological, and hydrological characteristics of the watershed. Precipitation and other meteorological data are the primary driving functions in the HSPF model. Surface runoff, stream flows, nonpoint source loads, and kinetic reaction rates all primarily depend on the continuous hourly input of precipitation, air temperature, evaporation, and other meteorological inputs.

Model calibration involves comparing the model results with observed data and adjusting key parameters to improve the accuracy of the model results. An acceptable model calibration requires a period long enough (usually several years) to reproduce different hydrologic conditions.

4.5   Hydrology Calibration

Hydrology calibration of the model compares simulated stream flow data to observed data. The model assumptions for hydrology are adjusted within reasonable ranges to achieve a good agreement in the comparison.

A comparison of the simulated and observed flow data indicates that the model calibration is robust and adequately reproduces the hydrologic response of the Four Mile Run watershed. There is a very good agreement between observed and simulated flow.

Because precipitation can vary across the watershed by 10 to 50 percent or more for any given storm, it is not realistic to expect simulated peak flows to match exactly with observed values. What is important is that the overall water balance is accurately reflected in total and seasonal flow volumes, and that error is minimized across the entire flow regime from drought conditions to infrequent storm events.

The TMDL model simulated baseflow adequately overall, with certain periods matching against gauged flows better than others. Since baseflows in Four Mile Run typically range from 2 to 10 cfs (quite low when compared to most other streams for which TMDL models are developed), even a few cfs difference can cause a model to appear significantly out of line when the response is quite good. Also, the USGS gauge site in Shirlington is in a very broad, shallow channel with an uneven, and ever-shifting, bottom. This makes developing and maintaining a rating curve for low flow and drought conditions a challenge. Thus, gauge error can account for some of the discrepancy between observed and simulated values during dry periods.

The most meaningful visual assessment of a model's accuracy across the entire range of flow conditions is seen in Figure 4, the flow-duration curve. For this curve, hourly flows were selected to increase the size of the dataset being analyzed, which adds resolution and results in smoother data plots. For this reason, Figure 4 shows that some simulated and observed hourly flows were in excess of 1,000 cfs. The X-axis in the flow-duration curve is deliberately stretched at the extremes of both low and high flows, to allow better assessment of the model's response to infrequent conditions. While simulated flows closely matched observed flows during storms of all sizes, as well as typical baseflow conditions, there is not a good agreement for the lowest half-percent of flows (about five days). This is an artifact of the model's start-up period. When the flow-duration curve is plotted for the period from January 7, 1999 to May 31, 2001, this outlier is removed. In reality, it is a difference of one to two cfs during the driest five days of the modeled period.

4.6   Summary of Key Hydrology Model Parameters Adjusted in Calibration

The primary parameters adjusted during the calibration were the infiltration capacity (INFLT), the recession rate for groundwater (AGWRC) the recession rate for interflow (IRC), the amount of evapotranspiration from the root zone (LZTEP), the amount of interception storage (CEPSC), and the amount of soil moisture storage in the upper zone (UZSN) and the lower zone (LZSN).

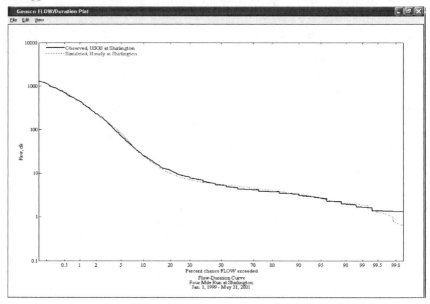

**Figure 4. Flow Duration Curve for Simulated and Observed
Hourly Flow at Shirlington**

4.7   Water Quality Modeling Approach - Source Representation

This section describes the approach taken for modeling the fate and transport of fecal coliform in Four Mile Run. The water quality portion of the model involved a linked two-step simulation process. First, the model simulated the fecal coliform concentration associated with the runoff (PQAL module of the PERLND section). Then, this load was transported in the different reaches using the GQAL module of the RCHRES section.

The PQAL module of HSPF was used to simulate the fecal coliform wash-off from the different land uses. The QUALOF option of PQAL was used to simulate the accumulation and removal of fecal coliform from the land by overland flow.

Next, the total fecal coliform loads for each source animal type were distributed over each of the land use categories that it occupies. Each animal type was evenly distributed over each of the land use categories that it occupies and the total fecal coliform loads for each animal type are spread evenly over the land use on a per acre basis. Table 4 shows the fecal coliform bacteria loading rate assumptions used for each species modeled and provides references for each assumption used.

Table 4. Modeled Fecal Coliform Bacteria Loading Rates by Host Species

Host Species	Fecal Coliform Production (count/animal/day)	Reference:
Waterfowl	7.99E+08	Canada Goose values from Accotink Creek TMDL, North River TMDL
Raccoon	4.09E+09	Best professional judgment
Human	1.88E+11	Mara & Oragui, 1981 (septic system equivalent)
Dog	4.09E+09	Long Island Regional Planning Board, 1978
Deer	5.00E+08	Interpolated from Metcalf & Eddy, 1991
Other Wildlife	1.88E+08	Average of four literature values for chicken

Table 5 shows the animal population densities by land use that were used for pervious segments (PERLNDs) in the TMDL model. These land use-specific population densities were arrived at with the aid of a spreadsheet through an iterative process to mimic daily bacteria loadings in proportion to the DNA evidence discussed in Chapter 3, as refined by interviews from the five naturalists. That is, while bacteria production rates for each animal were held constant using the values presented in Table 4, population densities for each animal were varied by land use in order to produce bacteria loads in proportion to the DNA evidence.

For pervious areas, daily bacteria loading rates for each animal source by land use were obtained by multiplying the animal densities presented in Table 5 by the daily fecal coliform bacteria production rates presented in Table 4. The actual daily bacteria loading rates for each PERLND used in the model were obtained by summing the loading rates for each animal source.

Table 5. Modeled Animal Densities by Land Use

Land Use	Density/acre[1]					
	Waterfowl	Raccoon	Human[2]	Dog[3]	Deer	Other[4] Wildlife
Open Space/Parks	6.0	0.45	0.0007	0.12	3.0	8.0
Highway	0.5	1.0	0.0008	0.3	0	5.0
Medium to High Density Mixed Use	3.0	1.0	0.03	0.4	0	3.5
Medium to High Density Industrial	2.2	0.9	0.03	0.27	0.2	10.0
Public/Conservation/Golf	6.0	0.45	0.0007	0.12	3.0	8.0
High Density Residential	4.1	0.5	0.019	0.25	0.2	3.0
Medium Density Residential	4.0	0.48	0.0095	0.32	1.2	7.0
Medium to High Density Residential	3.0	0.45	0.021	0.2	0.2	2.0
Medium to High Density Commercial	3.0	0.45	0.024	0.12	0	2.6
Low to Medium Density Residential	3.3	0.48	0.0028	0.62	1.2	8.4
Low Density Commercial	4.5	0.65	0.016	0.13	0.4	8.0
Low Density Industrial	4.5	0.52	0.016	0.22	0.6	8.0
Low Density Mixed Use	4.0	0.48	0.01	0.32	1.2	7.0
Federal	4.5	0.65	0.016	0.13	0.4	8.0

[1] Density values reflect the best professional judgment from a combination of factors, including in-stream DNA matches, long-term field observations, and adjustments to account for differing bacteria die-off rates among host species.
[2] Human population density reflects contributions from only sanitary sewer cross-connections and homeless assuming a per-capita septic system equivalent load.
[3] Dog densities reflect "non-picked-up" population only
[4] Other wildlife densities as estimated in equivalent chickens

For impervious segments (IMPLNDs) in the model, daily bacteria loads were obtained simply by taking each PERLND daily loading rate and dividing by a factor of 33. This factor is identical to that used in the Accotink Creek TMDL model (USGS, 2002, unpublished data). Unfortunately, this is an area for which very little research is available to guide the TMDL modeler. Although it seems intuitive that bacteria loading rates should be lower on impervious surfaces than pervious surfaces, there are no literature values to guide the selection of an impervious bacteria loading rate for different animals. This is because most studies have focused on impacts from livestock where impervious surfaces are not an issue. Bannerman (1993) and MS4 data from Arlington County (2001) have shown, however, that whatever the loading rates, fecal coliform bacteria counts from impervious surfaces are often in the tens of thousands colony-forming units (cfu) per 100 mL of water from stormwater runoff.

4.8 Existing Scenario Conditions

The water quality calibration runs were performed using the existing condition scenario. The intent of this scenario is to reproduce the long-term average fecal coliform fate and transport in the watershed. The simulation period selected for the calibration is from January 1, 1999 to May 31, 2001, which is the same as the hydrology calibration period. Bacteria calibration by matching simulated output to observed values is constrained by the following:

The model generates a daily mean value, but observed data are from instantaneous grab samples. Bacteria data is notoriously variable, and often fluctuates by an order of magnitude over the course of a day, even during seemingly static baseflow conditions (Gregory, 2001).

Observed data is often constrained by upper and lower detection limits. For example, of the 11 observed fecal coliform values collected by VADEQ in the model's calibration dataset, three are at a lower detection limit of 100, one is at a lower detection limit of 25, and one is at an upper detection limit of 8,000.

Nearly all of the bacteria data were collected during baseflow periods. Only one storm was chased for collection of fecal coliform data, and this was for NVRC's BST study, which used 1,600 cfu/100mL as its upper detection limit. All the samples collected during this storm (from July 14, 2000) were at this upper detection limit.

4.8.1 Water Quality Parameters

Several variables in the water quality model affect the simulation of the amount of fecal coliform washed off the land and transported through the Four Mile Run sub-watersheds. The most important variables are discussed below.

Rate of Surface Runoff That Removes 90 Percent of Stored Fecal Coliform Per Hour

One of the key parameters in the PQAL section that drives the amount of fecal coliform washed off the land is the rate of surface runoff that will remove 90 percent of stored fecal coliform per hour (WSQOP). WSQOP measures the susceptibility of the fecal coliform to wash off and adjusting it will change the fecal coliform peak concentrations during storm events. The final value used for the calibration is 2.0 inches per hour for pervious areas and 0.2 inches per hour for impervious areas, reflecting the reality that runoff from impervious surfaces occurs much more readily than runoff from pervious surfaces.

First Order Decay Rates of Fecal Coliform

Die-off from the pervious portions of the watershed was modeled with HSPF's first-order decay function. For all general quality constituents, the REMQOP factor is approximately equal to the first order decay coefficient, k. Thelin and Gifford (J. Environ. Qual. 12(1): 57-63) empirically determined this coefficient to be 0.11. Since REMQOP = ACQOP/SQOLIM, SQOLIM can be expressed as a multiple of ACQOP. Thus, the multiplication factor (MF) is the inverse of k=0.11, or 9, which was the peak summertime value used in the Four Mile Run model for each PERLND. This MF was varied monthly to account for observed seasonal differences in die-off noted in Section 2.3.1. The MF ranged from a low of 6.5 in January and February to a high of 9.0 in July and August, and is controlled by the monthly inputs for SQOLIM.

Impervious portions of the watershed also used the first order decay function. In research conducted by Olivieri et al, 1977, bacteria concentrations in urban streams was independent of the days since the last rainfall event, indicating either a very rapid buildup or an accumulation limit (maximum loading) not much greater than daily loading. Thus, a lower multiplication factor is expected for IMPLNDs than for PERLNDs, and an MF of 4 was arrived at through calibration.

In-stream die-off was also included in the model for which FSTDEC was set equal to 1.0. The transport of fecal coliform in model reaches uses the GQAL section of the RCHRES module. The key input parameter for the GQAL section is first order in-stream decay of fecal coliform. The value used in the calibration is at the low end of the published range of one to five and one half/day (Thomann, 1987) to reflect the limited in-stream bacteria die-off when compared with more pristine streams. However, this variable was not sensitive to the final simulated fecal coliform concentrations in the stream.

4.8.2   Results of the Water Quality Calibration

This section presents the analysis of the calibration results and discusses the main fecal coliform component loads in Four Mile Run. The calibrated model runs identify the major sources and their potential impact on the development of allocation scenarios. The model was run for the period from January 1999 to May 2001. Figures 5 and 6 show the results of the final water quality calibration run. These figures indicate reasonably good agreement between observed and simulated values.

As mentioned at the beginning of Section 4.8, one of the main reasons for wide discrepancies between simulated and observed bacteria values is that field measurements of bacteria are nearly always instantaneous grab samples, which can be highly variable across the course of each day, whereas simulated values are computed as daily averages. This is shown in Figures 5 and 6 where some of the observed-instantaneous fecal coliform values differ from their corresponding simulated values. Also, it is likely that had the observed data that was constrained by the upper and lower detection limits been allowed to reflect accurate readings, a somewhat better fit would have been demonstrated. Overall, however, the model used for this TMDL captures the range of observed values sufficiently well.

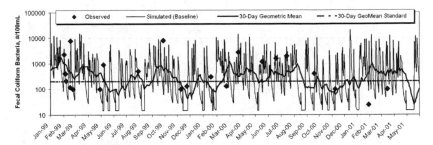

Figure 5. Simulated and Observed Daily Fecal Coliform, Log Scale

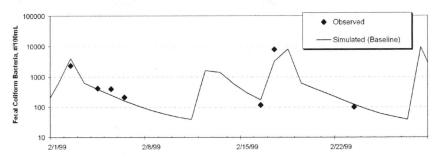

**Figure 6. Sample Detail of Simulated and Observed Daily Fecal
Coliform, Log Scale**

5. Load Allocations

5.1 Background

The objective of a TMDL plan is to allocate allowable loads among the various
pollutant sources so that the appropriate control actions can be taken to achieve water
quality standards. The specific objective of the TMDL plan in Four Mile Run is to
determine the required reductions in fecal coliform loadings from various non-point
sources in order to meet state water quality standards. The state water quality
standard for fecal coliform used in the TMDL development is the 30-day geometric
mean of 200 counts/100 mL. The incorporation of the different sources into the
TMDL is defined in the following equation (USEPA, 1999):

TMDL = WLA + LA + MOS
Where:
WLA = waste load allocation (point sources)
LA = load allocation (non-point sources)
MOS = margin of safety

The allocation scenario for Four Mile Run was designed to meet the water quality
standard of a geometric mean of 200 counts/100 mL. An MOS of 5 percent was
incorporated explicitly in the TMDL equation by reducing the target fecal coliform

concentration from 200 counts/100 mL to 190 counts/100 mL. In other words, the simulated concentrations were compared to a target of a geometric mean (of 30 data points) of 190 counts/100 mL. The time period selected for the load allocation covers the same period used in the water quality calibration (January 1999 to May 2001) and it includes both high and low flow conditions. The results of the simulation for the existing conditions are presented in Section 5.5.3.

5.2　Allocations Scenarios

The TMDL development requires that the level of reduction from each pollutant in a watershed be determined in order to meet the applicable water quality standard. The TMDL comprises the sum of individual waste load allocations (WLAs) for point sources and load allocations (LAS) for non-point sources. However, as explained in the following section, there are no WLAs for fecal coliform bacteria in the nontidal portion of the Four Mile Run watershed.

5.2.1　Wasteload Allocations

There are no VPDES permits that allow discharge of fecal coliform from point sources to the nontidal portion of Four Mile Run. Arlington County's 30 million gallon/day sewage treatment plant discharges downstream of the tidal/non-tidal boundary of this TMDL and easily complies with its 200 counts/100 mL limits specified in its VPDES permit. However, because the counties of Arlington and Fairfax have existing municipal separate storm sewer (MS4) permits, and because Alexandria and Falls Church are expected to receive MS4 permits in the near future, wasteload allocations (WLAs) for this TMDL were developed based on contributions from impervious surfaces in the study area. The basis for these impervious contributions is explained in Section 4.7.

5.2.2　Load Allocations

Four load allocation scenarios were evaluated to meet the TMDL goal of a 30-day geometric mean of 190 counts/100 mL. These scenarios are summarized in Table 6.

Table 6. Existing Conditions and TMDL Allocation Scenarios for Four Mile Run

	Reduction in Loadings from Existing Conditions (%)					% days Geometric Mean > than 190 MPN/100mL
	Waterfowl	Raccoon	Human	Dog	Other Wildlife	
Existing Conditions	0	0	0	0	0	65
Scenario 1	0	0	95	95	0	54
Scenario 2	50	50	95	95	0	41
Scenario 3	80	80	98	98	80	8
Scenario 4	95	95	98	98	95	0

Scenario 1 assesses the fecal coliform contribution of wildlife to Four Mile Run, with a 95% reduction in loadings from humans and dogs. The objective of this initial scenario is to assess the possibility of developing a TMDL allocation plan that meets state water quality standards only by reducing sources of fecal coliform caused by human activities, including management of pet waste. Scenario 1 indicates that the fecal coliform due to wildlife causes concentrations in the stream to violate the 30-day geometric mean 54% of the time. This scenario indicates that eliminating load allocations of fecal coliform caused by human activities (including controlling 95% of the pet waste) will not provide a TMDL that meets the Virginia water quality standards.

Scenario 2 assesses the impact of reducing by 95% the direct sources from human activities (including pet waste) and a 50% reduction in anthropogenic wildlife (resident urban waterfowl and raccoons). Under this scenario the 30-day geometric mean, with the margin of safety, is exceeded 41 percent of the time, which indicates that further load reductions are needed.

Scenario 3 examines the benefits of reducing fecal coliform bacteria from all wildlife sources by 80% and from humans and dogs by 98%. Under this scenario, bacteria counts are expected to exceed the 190 TMDL limit eight percent of the time.

Scenario 4 is the only modeled scenario that is demonstrated to achieve the goals of the TMDL. It considers the case of controlling 98% of the fecal coliform bacteria from humans and dogs, as well as 95% of the bacteria from all wildlife. Loadings from this scenario for each land use serve as the basis for the numbers in the final TMDL shown in Table 7.

Table 7. Annual Fecal Coliform Loadings (counts/year) Used for Developing the Fecal Coliform TMDL for Four Mile Run

Parameter	WLA	LA	MOS*	TMDL
Fecal coliform	2.04E+13	9.61E+14	4.91E+13	1.03E+15

* Five percent of the TMDL

5.3   Future Growth

Although the Four Mile Run watershed is virtually built out in terms of existing land use reflecting current land use plans, the potential exists for small additions of infill development and population expansion. Census data shows that despite being nearly built out, population has increased steadily over the past several decades. For instance, NVRC's analysis of new census data shows an increase of nearly 11 percent from a population of 165,000 in 1990 to 183,000 in 2000. The pet population has almost certainly increased as well, although probably by less than 11 percent, as the majority of newer residents live in multi-family dwellings where pet ownership is restricted and many are recent immigrants that come from cultures with less of a tradition of owning pets. Further, some anthropogenic wildlife species, like resident geese and raccoons, have increased their numbers in the face of urbanization (Hadidian, 1997 and 1991). As a result of the intense development pressures in this watershed, driven largely by infill opportunities, there is reason to suspect that urban wildlife populations may have approached their carrying capacity locally.

The assumptions used in the model to develop estimates of fecal coliform loads are conservative and provide for a reasonable assurance that the estimated loads account for changes in the land use and populations in the Four Mile Run watershed.

5.3.1 Consideration of Critical Conditions

EPA regulations at 40 CFR 130.7 (c)(1) require TMDLs to take into account critical conditions for stream flow, loading, and water quality parameters. The intent of this requirement is to ensure that the water quality of Four Mile Run is protected during times when it is most vulnerable. Critical conditions are important because they describe the factors that combine to cause a violation of water quality standards and will help in identifying the actions that may have to be undertaken to meet water quality standards.

The sources of bacteria for Four Mile Run were a mixture of dry and wet weather driven sources. TMDL development utilized a continuous simulation model that applies to both high and low flow conditions. Consequently, the critical conditions for Four Mile Run were addressed during TMDL development.

5.4 TMDL Implementation

DEQ intends for this TMDL to be implemented through best management practices (BMPs) in the watershed. Implementation will occur in stages. The benefits of staged implementation are:

1. as stream monitoring continues to occur, it allows for water quality improvements to be recorded as they are being achieved;

2. it provides a measure of quality control, given the uncertainties which exist in any model;

3. it provides a mechanism for developing public support;

4. it helps to ensure the most cost effective practices are implemented initially; and

5. it allows for the evaluation of the adequacy of the TMDL in achieving the water quality standard.

If a staged approach to implementation were followed, a useful interim reduction goal would be to achieve an instantaneous standards violation rate of 10% or less, because under the current monitoring frequency, this would allow Four Mile Run to be removed from the 303d impaired waters list. The scenarios shown in Tables 6 and 7 offer one approach to staging bacteria reductions. Table 8 shows the percent of days that the TMDL model predicts will violate the instantaneous standard for fecal coliform of 1000 MPN/100 mL. This table shows that the instantaneous standard will be met 90% of the time with a scenario that is intermediate of Scenarios 2 and 3, thus achieving this interim reduction goal.

**Table 8. Existing Conditions and TMDL Allocation Scenarios
for Staged Implementation**

	Reduction in Loadings from Existing Conditions (%)					% days > than 1000 MPN/100mL
	Waterfowl	Raccoon	Human	Dog	Other Wildlife	
Existing Conditions	0	0	0	0	0	24
Scenario 1	0	0	95	95	0	17
Scenario 2	50	50	95	95	0	13
Scenario 3	80	80	98	98	80	4
Scenario 4	95	95	98	98	95	0.1

In general, the Commonwealth intends for the required reductions to be implemented in an iterative process that first addresses those factors with the largest impact on water quality. For example in urban area like the Four Mile Run watershed, reducing the human bacteria loading from damaged or cross-connected sanitary sewer lines could be a focus during the first stage because of its health implications. This component could be implemented through stepped-up sanitary sewer inspections and sewer rehabilitation programs. Other management practices that might be appropriate for controlling urban wash-off from parking lots and roads and that could be readily implemented may include high efficiency street sweeping, improved garbage collection and control, and increasing the number of dog parks and improving their siting and management. Many of these practices have already been initiated and are being implemented in some of the local jurisdictions that share the watershed.

Adding and retrofitting regional ponds, such as those suggested in a report on the feasibility of regional ponds in the Four Mile Run watershed (Northern Virginia Planning District Commission, 1993), has the potential to improve water quality on multiple fronts. It is worth exploring the idea that fecal coliform levels downstream of such facilities may be partially mitigated by designing the pond outlet to release from an optimized depth less affected by bacteria on the water surface or in the sediments.

6. Reasonable Assurance for Implementation

6.1 Follow-Up Monitoring

The Department of Environmental Quality will continue to monitor Four Mile Run in accordance with its ambient monitoring program. VADEQ and VADCR will continue to use data from these monitoring stations to evaluate reductions in fecal bacteria counts and the effectiveness of the TMDL in attaining and maintaining water quality standards.

6.2 Regulatory Framework

This TMDL is the first step toward the expeditious attainment of water quality standards. The second step will be to develop a TMDL implementation plan, and the final step is to implement the TMDL until water quality standards are attained.

Watershed stakeholders will have opportunities to provide input and to participate in the development of the implementation plan, which will also be supported by regional and local offices of VADEQ, VADCR, and other cooperating agencies.

7. Public Participation

The development of the Four Mile Run TMDL would not have been possible without public participation. The first public meeting was held in Arlington on June 14, 2001 to discuss the water quality data and development of the TMDL. About 25 people attended. Copies of the presentation materials and diagrams outlining the development of the TMDL were available for public distribution. A public notice was placed in the *Virginia Register* about this meeting and a 30 day-public comment period. Four written public comments were received. A second public notice was published in the *Virginia Register* on March 11, 2002 to advertise a second public meeting in Alexandria on March 25, 2002 and a 30 day-public comment period ended on April 9.

Two themes emerged from the first round of comments. One was a desire to increase baseflow to the stream as a means for diluting bacteria levels and to begin to restore more natural background levels of bacteria. There was a desire to see micro-drainage, infiltration BMPs implemented in the watershed in a significant way.

The second theme mentioned was a strong caution against attempting to change the current designated use of Four Mile Run as a stream used for primary contact recreation. While four voices from within a watershed population of 183,000 is not a consensus, and may not be consistent with the desires of some local government staff, the point was made that Four Mile Run is regularly used for contact recreation primarily because of its sheer proximity to a large urban population and its excellent public access through its greenway park system and popular streamside trails.

Many valuable inputs were received during the second round of comments, and a number have been addressed in the changes made between the draft and final TMDL report. These comments helped make a stronger, more useful TMDL document all-around.

8. References

1.   Abugattas, Alonzo (Arlington County naturalist, Long Branch Nature Center), August 30, 2001. Personal communication (face-to-face interview).
2.   American Society of Agricultural Engineers (ASAE). 1998. *ASAE Standards, 45th edition: Standards, Engineering Practices, Data.* St. Joseph, MI.
3.   Arlington County. 2001. *Watershed Management Plan.*
4.   Bannerman, R., D. Owens, R. Dodds, and N. Hornewer. 1993. Sources of Pollutants in Wisconsin Stormwater. *Water Science and Technology*, 28(3-5):241-259.
5.   Beaudeau, P., N. Tousset, F. Bruchon, A. Lefevre, and H. Taylor. 2001. In situ measurement and statistical modeling of Escherichia Coli decay in small rivers. *J. Wat. Res.*(Elsevier Science Ltd.), v. 61. 35, #13, pp. 3168-3178.
6.   Chauvette, Denise (Arlington County naturalist, Gulf Branch Nature Center), August 28, 2001. Personal communication (face-to-face interview).

7.   Davies, C.M., J.A.H. Long, M. Donald, and N.J. Ashbolt. 1995. Survival of
      fecal microorganisms in marine and freshwater sediments. *Appl. Environ.
      Microbiol.* 61: 1888-1896.
8.   Deibler, Scott (Arlington County naturalist, Gulf Branch Nature Center),
      August 28, 2001. Personal communication (face-to-face interview).
9.   Environmental Systems Analysis, Inc. 1999. Appendix C – Baseline
      Monitoring (part of a survey of macrobenthic diversity in Four Mile Run).
10.   Frost, W.H. 1998. Precipitation Analysis for Washington, D.C., unpublished
      report from Arlington County Department of Public Works.
11.   Gerba, C.P. and J.S. McLoed. 1976. Effect of sediments on the survival of
      Escherichia coli in marine waters. *Appl. Environ. Microbiol.* 32: 114-120.
12.   Geldreich, E.E. 1978. Bacterial populations and indicator concepts in feces,
      sewage, stormwater and solid wastes. In Indicators of Viruses in Water and
      Food, ed. G. Berg, ch. 4, 51-97. Ann Arbor, Mich.: Ann Arbor Science
      Publishers, Inc.
13.   Gregory, M. Brian and Elizabeth A. Frick (USGS). 2001. *Indicator Bacteria
      Concentrations in Streams of the Chattahootchee River National Recreation
      Area, March 1999 – April 2000.* Proceedings of the 2001 Georgia Water
      Resources Conference, held March 26-27, 2001.
14.   Hadidian, J, D.A. Manski, and S. Riley. 1991. *Daytime resting site selection in
      an urban raccoon population*, pp. 39-45. In LW Adams and DL Leedy (eds),
      Wildlife Conservation in Metropolitan Environments . National Institute for
      Urban Wildlife, Symposium, Ser.2, 10921 Trotting Ridge Way, Columbia, MD,
      21044.
15.   Hadidian, J., G.R. Hodge, and J.W. Grandy (eds). 1997. *Wild Neighbors*.
      Humane Society of the United States, Fulcrum Publishing, 350 Indiana Street,
      Suite 350, Golden, Colorado 80401- 5093.
16.   Horner, R.R. 1992. Water quality criteria/pollutant loading
      estimation/treatment effectiveness estimation. In R.W. Beck and Associates.
      Covington Master Drainage Plan. King County Surface Water Management
      Division. Seattle, WA.
17.   Horsley & Whitten. 1996. Identification and Evaluation of Nutrient and
      Bacteriological Loadings to Maquoit Bay, Brunswick, and Freeport, Maine.
      Final Report. Casco Bay Estuary Project, Portland, ME.
18.   Leeming, R., N. Bate, R. Hewlett, and P.D. Nichols. 1998. Discriminating
      Faecal Pollution: A Case Study of Stormwater Entering Port Phillip Bay,
      Australia. Accepted for publication in J. Water Science and Technology.
19.   Long Island Regional Planning Board (LIRPB). 1978. *The Long Island
      Comprehensive Waste Treatment Management Plan: Volume II: Summary
      Documentation.* Nassau-Suffolk Regional Planning Board. Hauppauge, NY.
20.   Lumb, A.M. and J.L. Kittle, Jr. 1993. Expert system for calibration and
      application of watershed *models. In Proceedings of the Federal Interagency
      Workshop on Hydrologic Modeling Demands for the 90's*, ed. J.S. Burton.
      USGS Water Resources Investigation Report 93-4018.
21.   Maptech. 2000. Fecal Coliform TMDL for the Middle Blackwater River,
      Virginia.

22. Mara, D.D. and J.I. Oragui. 1981. Occurrence of *Rhodococcus coprophilus* and associated actinomycetes in feces, sewage, and freshwater, *Appl. Environ. Microbiol.* 42: 1037-42.

23. Marino, R.. and J. Gannon.1991. Survival of fecal coliforms and fecal streptocci in storm drain sediment. *Water Resources* 25(9): 1089-1098.

24. Metcalf & Eddy. 1991. *Wastewater Engineering: Treatment, Disposal and Reuse.* Third edition. George Tchobanoglous and Franklin L. Burton, Eds.

25. Doug Moyers. 2000-2002. USGS project manager for the Accotink Creek TMDL model. *Personal Communication (various occasions).*

26. Murphy, D.D. 1988. *Challenges to biological diversity in urban areas,* pp. 71-76. In EO Wilson and FM Peter (eds), Biodiversity. National Academy Press, Washington, DC.

27. National Weather Service. 2002. Local Climatological Data: 2001 Annual Summary with Comparative Data, Washington, D.C., Ronald Reagan National Airport (DCA), ISSN 0198-1196.

28. Northern Virginia Planning District Commission. 1996. *Four Mile Run Watershed In-stream Water Quality Final Report.* Annandale, Virginia 22003.

29. Northern Virginia Planning District Commission. 1994. *Dog Waste Contributions to Urban NPS Pollution* (unpublished white paper). Annandale, Virginia 22003.

30. Northern Virginia Planning District Commission. 1994. *Regional BMPs in the Four Mile Run Watershed, A Feasibility Investigation.* Annandale, Virginia 22003.

31. Northern Virginia Regional Commission. 2000. *Optical Brightener Monitoring in the Four Mile Run Watershed,* abstract in Virginia Water Resources Research Symposium, Virginia Tech, Roanoke, Virginia (November 2000). Annandale, Virginia 22003.

32. Northern Virginia Regional Commission. 2001. *Staff Analysis of 2000 U.S. Census data* (unpublished). Annandale, Virginia 22003.

33. Ogle, Martin (Chief Naturalist for Potomac Overlook Regional Park, Northern Virginia Regional Park Authority), September 12, 2001. Personal communication (face-to-face interview).

34. Olivieri, V., C. Kruse, K. Kawata, and J. Smith. 1977. *Microorganisms in Urban Stormwater.* USEPA Report No. EPA-600/2-77-087 (NTIS No. PB-272245).

35. Riley, S., J. Hadidian, and D.A. Manski. 1998. Population density, survival, and rabies in raccoons in an urban national park. *Can. J. Zool.* 76:1153-1164.

36. SAIC. 2001. Fecal Coliform TMDL for Holmans Creek, Virginia.

37. Simmons, G.M., Jr., and D.F. Waye. 2001. Estimating Nonpoint Fecal Coliform Sources in Northern Virginia's Four Mile Run Watershed, *Advances in Water Monitoring Research,* edited by Tamim Younos., Water Resources Publications, ISBN 1-887201-33-5.

38. Simmons, G.M., Jr. 1994. Potential sources for nonpoint introduction of Escherichia coli (E. coli) to tidal inlets. *Interstate Seafood Conference, Proceedings.* Rehobeth Beach, Delaware.

39. Simmons, G.M., Jr., S.A. Herbein, and C.A. James. 1995. *Managing nonpoint fecal coliform sources to tidal inlets*. Water Res. Update. Issue 100: 64-74.
40. Stephenson, G.R. and R.C. Rychert. 1982. Bottom sediment: a reservoir of Escherichia coli in rangeland streams. Jour. Range Management 35: 119-124.
41. Sherer, B.M., J.R. Miner, J.A. Moore, and J.C. Buckhouse. 1992. Indicator bacterial survival in stream sediments. *J. Environ. Qual.* 21: 591-595.
42. State Water Control Board. 1997. *Water Quality Standards*. Effective date, December 10, 1997.
43. Thelin, R. and G. F. Gifford. 1985. Fecal coliform release patterns from fecal material of cattle. J. Environ. Qual. 12(1):57-63.
44. Thomann, R. 1987. *Principles of Surface Water Quality Modeling and Control*. Harper and Row, Publishers, New York.
45. USEPA. 2001. *Protocol for Developing Pathogen TMDLs*. EPA 841-R-00-002. Office of Water (4503F), Washington, DC. 132 pp.
46. USEPA. 1999. Guidance for Water Quality-Based Decisions: The TMDL Process.
47. USGS. 2001. Unpublished data (Accotink Creek model inputs).
48. Virginia Department of Environmental Quality (DEQ). 2000. *305(b) Report to the EPA Administrator and Congress for the Period January 1, 1994 to December 31, 1998*. Department of Environmental Quality and Department of Conservation and Recreation. Richmond, Virginia.
49. Virginia Department of Environmental Quality (DEQ) and Virginia Department of Conservation and Recreation (DCR). 1998. *303(d) Total Maximum Daily Load priority list report*. Richmond, Virginia.
50. Virginia Department of Environmental Quality (DEQ). 1996. *Virginia Water Quality Assessment for 1996 and Non-Point Source Watershed Assessment Report*. Department of Environmental Quality and Department of Conservation and Recreation. Richmond, Virginia.
51. Virginia Tech Department of Biology. 2000. *Fecal Coliform TMDL for Pleasant Run*.
52. Yagow, Eugene. 2001. Virginia Tech, Fecal Coliform TMDL for Mountain Run, Virginia.
53. Zell, Greg (Arlington County naturalist, Long Branch Nature Center), August 30, 2001. Personal communication (face-to-face interview)

Total Maximum Daily Load for Nitrogen Compounds for the Los Angeles River and its Tributaries

Brian Watson[1], Drew Ackerman[2], Ken Schiff[3], and Terrence Fleming[4]

[1] Tetra Tech, Inc., 2110 Powers Ferry Road, Suite 202, Atlanta, Georgia 30339
(770) 850-0949
[2] Southern California Coastal Water Research Project, 7171 Fenwick Lane, Westminster,
California 92683, (714) 372-9217, www.swrcb.ca.gov
[3] Southern California Coastal Water Research Project, 7171 Fenwick Lane, Westminster,
California 92683, (714) 372-9209, www.swrcb.ca.gov
[4] U.S. EPA Region 9, 75 Hawthorne St. (WTR-2), San Francisco, California 94105
(415) 947-8000

Abstract

The State of California is required to develop Total Maximum Daily Loads (TMDLs)
for waters not meeting water quality standards, in accordance with Section 303(d) of
the Clean Water Act and the U. S. Environmental Protection Agency (EPA) Water
Quality Planning and Management Regulations (40 CFR Part 130). The TMDLs
developed for LA River and its tributaries consists of segments impaired by nitrogen
compounds and their related instream eutrophic effects (low dissolved oxygen, low
pH and excessive algae). This report is a summary of the TMDLs for nitrogen
compounds and related eutrophic effects within the LA River and its tributaries.
The LA River watershed is one of the largest in the region, covering 819 square miles
(mi^2). It is also one of the most diverse in terms of land use patterns. There are
several potential sources of pollutants in the watershed.

To determine the sources of impairments, several tools were used. Due to the many
flow related factors that influence in-stream nutrient, algal, and dissolved oxygen
concentrations, a hydrodynamic model linked with a water quality model was needed.
The 1-dimensional version of the hydrodynamic model Environmental Fluid
Dynamics Code (EFDC) linked with the Water Quality Analysis Simulation Program
(WASP) water quality model were selected for the LA River application.

The TMDL sets wasteload allocations (WLAs) for ammonia, nitrate, and nitrate +
nitite for the three major POTWs in the watershed and other NPDES dischargers.
The WLAs will be established as NPDES permit effluent limits.

TABLE OF CONTENTS

LIST OF TABLES

LIST OF FIGURES

1. Introduction

The State of California is required to develop Total Maximum Daily Loads (TMDLs) for waters not meeting water quality standards, in accordance with Section 303(d) of the Clean Water Act and the U. S. Environmental Protection Agency (EPA) Water Quality Planning and Management Regulations (40 CFR Part 130). Several segments of the Los Angeles (LA) River and its tributaries were included on EPA's 303(d) list of impaired waters in California for a variety of pollutants. A Consent Decree established a schedule of development for TMDLs in the LA Region and grouped the 700 waterbody-pollutant combinations into 92 analytical units for TMDL development. The TMDLs developed for LA River and its tributaries represent Analytical Unit 11 of the Consent Decree, which consists of segments impaired by nitrogen compounds and their related instream eutrophic effects (low dissolved oxygen, low pH and excessive algae). TMDLs established to address these impairments are presented in Los Angeles Regional Water Quality Control Board (2003).

This report is a summary of the TMDLs for nitrogen compounds and related eutrophic effects within the LA River and its tributaries. The report will contain many of the elements that are common to a TMDL report, including but not limited to water quality standards, source assessment, development of a system of models, TMDL scenarios, allocations, critical conditions, margin of safety and an implementation plan. Much of the information from this report is contained in further detail in two primary documents 1) Modeling Analysis for the Development of TMDLs for Nitrogen Compounds in the Los Angeles River and Tributaries (Tetra Tech, 2002) and 2) Total Maximum Daily Loads for Nitrogen Compounds and related effects – Los Angeles River and Tributaries (Los Angeles Regional Water Quality Control Board, 2003). For more information about the LA River Nitrogen TMDL, please refer to these documents. The work presented herein was performed in cooperation with EPA Region 9, the Southern California Coastal Water Research Project, the Los Angeles Regional Water Quality Control Board, the City of Los Angeles, the Los Angeles County Department of Public Works, and the Los Angeles and San Gabriel Rivers Watershed Council.

1.1 Study Area Description

The 55-mile LA River flows from the Santa Monica Mountains at the western end of the San Fernando Valley to the Pacific Ocean. The headwaters of the LA River are located in the Santa Monica Mountains at the confluence of Arroyo Calabasas and Bell Creek. Arroyo Calabasas drains Woodland Hills Calabasas, and Hidden Hills in the Santa Monica Mountains. Bell Creek drains the Simi Hills, and receives discharges from Chatsworth Creek. From the confluence of Arroyo Calabasas and Bell Creek, the LA River flows east through the southern portion of the San Fernando Valley, a heavily developed residential and commercial area. Major tributaries to the river in the San Fernando Valley are the Pacoima Wash, Tujunga Wash (both drain portions of the Angeles National Forest in the San Gabriel Mountains), Burbank

Western Channel, and Verdugo Wash (both drain the Verdugo Mountains). The LA River turns in an area known as the Glendale Narrows and flows south for approximately 25 miles through industrial and commercial areas and is bordered by railyards, freeways, and major commercial and government buildings. Below the Glendale Narrows, three major tributaries feed the LA River—Arroyo Seco Wash, Rio Hondo, and Compton Creek. The river discharges to the Pacific Ocean at Queensway Bay, a portion of San Pedro Bay in Long Beach. Figure 1 shows the LA River watershed in relation to neighboring counties and the State of California.

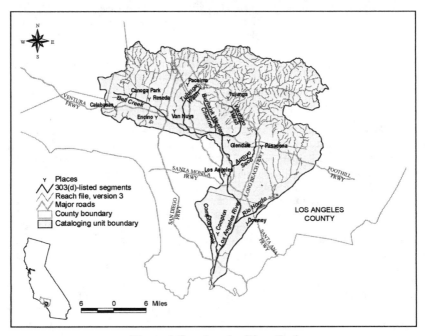

Figure 1. Los Angeles River Basin

Due to major flood events at the beginning of the century, most of LA River was lined with concrete by the 1950s. In the San Fernando Valley, there is a section of the river with a soft bottom at the Sepulveda Flood Control Basin, a 2,150-acre open space upstream of the Sepulveda Dam that is designed to collect flood waters during major storms. In the area around the Glendale Narrows, the water table was too high to allow laying of concrete; the river in this area has a rocky, unlined bottom with concrete-lined or rip-rap sides. This stretch of the river is fed by natural springs and supports stands of willows, sycamores, and cottonwoods. South of the Glendale Narrows, the river is contained in a concrete-lined channel down to Willow Street in Long Beach.

The Rio Hondo, through the Whittier Narrows Reservoir, hydraulically connects the river to the San Gabriel River Watershed. Flows from the San Gabriel River and Rio

Hondo merge at this reservoir during larger flood events, and flows from the San Gabriel River watershed may impact the LA River. Most of the water in the Rio Hondo is used for groundwater recharge during dry weather.

The LA River watershed is one of the largest in the region, covering 819 square miles (mi2). It is also one of the most diverse in terms of land use patterns. Figure 2 and Table 1 present the landuse distribution throughout the LA River watershed, based on 1994 data from the Los Angeles County Department of Public Works (LACDPW). Seven general landuse categories were used for the purposes of characterizing the watershed. Approximately 364 mi^2 of the watershed are covered by forest and open space mostly concentrated at the headwaters in the Santa Monica, Santa Susana, and San Gabriel Mountains. The remainder of the watershed is highly developed. Landuse patterns within the LA River Watershed closely follow the topographic features. The mountainous regions are primarily open forested land while the low-lying areas are a mixture of high-density residential, industrial and commercial uses.

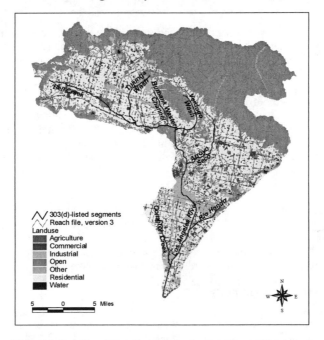

Figure 2. Landuse Distribution in the LA River Watershed

Table 1. Landuse Areas in the LA River Watershed

Landuse	Area (acres)	Area (mi^2)	Percentage of Total
Open	232,832	363.80	43.60%
Residential	189,645	296.32	35.51%
Industrial	55,377	86.53	10.37%
Commercial	39,878	62.31	7.47%
Agricultural	3,817	5.96	0.71%
Other	1,654	2.58	0.31%
Water	1,069	1.67	0.20%
Total Area	524,272	819.18	100.00%

The LA River has two distinct flow conditions as a result of the prevailing rainfall patters in the region. Typically the high-flow (or wet weather) conditions occur between October and March, while the low-flow (or dry weather) conditions occur from April through September. The wet weather periods are marked by events where flows in the river and tributaries rise and fall rapidly, reaching flow levels on the order of thousands of cubic feet per second (cfs). Flows during the wet weather periods are generated by storm runoff in the watershed. Stormwater runoff in the sewered urban areas of the watershed is carried to the river through a system of approximately 5,000 miles of storm drains. During times of higher flow, stormwater runoff delivers nutrients from nonpoint sources in the watershed. The stormwater also increases the volume of water in the river, creating a larger capacity for assimilating pollutant loads.

In between rainfall events and during low-flow periods, the flows are significantly lower and less variable. The predominant contribution to instream flow and nutrient loading comes from the primary point source discharges to the system. Discharges from the three major point sources (water reclamation plants) can comprise up to 80 to 100 percent of the flow in the LA River and over 80 percent of the nitrogen load.

1.2   Extent of Impairments

Regional Water Quality Control Boards (Regional Board) in California define beneficial uses for waterbodies in Water Quality Control Plans (Basin Plans). Numeric and narrative water quality objectives are specified in each region's Basin Plan to be protective of the beneficial uses in each waterbody in the region. The Los Angeles Regional Board Basin Plan defines 14 potential, intermittent, or existing beneficial uses for the LA River. Several segments of the LA River watershed (Analytical Unit 92) were included on EPA's 303(d) list of impaired waters in California for not supporting their uses due to a variety of pollutants, including pH, ammonia, metals, coliform, trash, scum, algae, oil, chlorpyrifos, pesticides, and volatile organics (Table 2 and Figure 3). Beneficial uses impaired by nitrogen compounds and related eutrophic effects (e.g., low dissolved oxygen and excessive algae) are those associated with aquatic life, wildlife habitat, recreation and groundwater recharge. The critical conditions for impairment occur during times of low flow when instream flow and nutrient loading are dominated by point source discharges. Therefore, too account for these critical conditions, the TMDLs developed for nitrogen compounds in LA River and its tributaries are developed for the critical low-flow period. The TMDLs were established to meet applicable water quality standards. Table 3 presents the numeric water quality targets derived from review of applicable standards and used in establishing the LA River TMDLs. More details about the establishment of TMDL targets to support beneficial uses and address impairments are included in the LA River TMDLs (Los Angeles Regional Water Quality Control Board, 2003).

Table 2. Segments in the LA River watershed listed for nutrients and related impairments

Listed Segment	Length of Impaired Segment by Pollutant (miles)					
	Ammonia	Nitrogen	Algae	Odors	Scum/ Foam	pH
Los Angeles River (at Sepulveda Basin)	1.9	1.9	-	1.9	1.9	-
Los Angeles River (from Sepulveda Dam to Riverside Dr.)	11.8	11.8	-	11.8	11.8	-
Los Angeles River (from Riverside Dr. to Figueroa St.)	7.2	7.2	-	7.2	7.2	-
Tujunga Wash (from Hansen Dam to Los Angeles River)	9.7	-	-	9.7	9.7	-
Burbank Western Channel	6.4	-	6.4	6.4	6.4	-
Verdugo Wash (from Verdugo Rd. to Los Angeles River)	-	-	3.4	-	-	-
Arroyo Seco (from West Holly Ave. to Los Angeles River)	-	-	7.0	-	-	-
Los Angeles River (from Figueroa St. to Carson St.)	19.4	19.4	-	19.4	19.4	-
Rio Hondo (at the Spreading Grounds)	2.7	-	-	-	-	-
Rio Hondo (from the Santa Ana Fwy. to Los Angeles River)	4.2	-	-	-	-	4.2
Compton Creek	-	-	-	-	-	8.5
Los Angeles River (from Carson St. to estuary)	2.0	2.0	-	2.0	2.0	-

Figure 3. 303(d)-listed Segments in the Los Angeles River Basin

Table 3. Numeric water quality targets for the LA River nutrient TMDLs (USEPA, 2002)

Parameter	Beneficial Use[1]	Numeric Target
Ammonia-N	WILD, WARM	Temperature and pH dependent: 4.0 mg/L acute; 2.0 mg/L chronic
Nitrate-N	(established in Basin Plan)	8 mg/L
Nitrite-N	GWR	1 mg/L
Nitrate-N+Nitrite-N	GWR	8 mg/L in the LA River and Rio Hondo; 10 mg/L in other tributaries
pH	WILD, WARM	6.5 to 8.5
Dissolved oxygen	WILD, WARM	Average of 7 mg/L, but not less than 5 mg/L
[1]WILD = Wildlife habitat; WARM = Warmwater habitat; GWR = Groundwater recharge		

2.   Technical Modeling Approach

When selecting an appropriate technical approach for a water quality study, it is important to identify and understand the defining characteristics of the waterbody system, the goals and planned use of the modeling system, and any unique aspects of the waterbody or impairment that will guide the approach. A technical committee comprised of representatives from various agencies coordinated the selection of an appropriate modeling approach for addressing the nitrogen and related impairments in the LA River and tributaries, as well as supporting monitoring. This committee included representatives from EPA Region 9, the Los Angeles Regional Water Quality Control Board, the Southern California Coastal Water Research Project, the City of Los Angeles, Los Angeles County Department of Public Works and the Los Angeles and San Gabriel Rivers Watershed Council. EPA contributed contract funds for the modeling. The other groups contributed either funds or in-kind services to monitoring and special studies designed to support the model development.

The following sections present the information that led to the selection of the technical approach and descriptions of the chosen models and their applicability to the evaluation of the LA River.

2.1   Guiding Assumptions

The LA River is a complex and unique system with many concrete-lined channels and distinct hydrologic behavior and responses, some of the major characteristics that define the evaluation of nitrogen compounds and related eutrophic effects in the river were identified prior to model selection. The following "guiding assumptions" represent factors that shaped the model selection and development for the LA River TMDLs for nitrogen compounds.

- The approach for TMDLs should evaluate the entire watershed, rather than take a reach-by-reach approach. The LA River watershed contains 12 segments listed for nitrogen compounds and related impairments, including both mainstem and tributary segments. Because many of the listed segments affect the conditions of downstream listed segments, it is important to be able to evaluate the relationship between the segments.
- The LA River should be simulated as a waterbody with all the potential riverine features, including hydrologic/hydraulic transport, drainage, and chemical and biological activity.
- The modeling approach for the LA River should be designed to allow for its use in future TMDLs for bacteria and metals in the river. In addition, to provide consistency throughout the region, it is anticipated that the LA River approach will also be used to develop other TMDLs in the region (e.g., San Gabriel River TMDLs).
- The LA River TMDLs focus on nitrogen compounds as well as their related eutrophic effects in a receiving waterbody (e.g., algal growth, low dissolved oxygen). Therefore, the model chosen must be capable of simulating these parameters, as well as other pollutants to be addressed in future TMDLs.
- The LA River experiences two distinct flow conditions associated with wet and dry weather. Although this TMDL focuses on low flow conditions, future applications and TMDLs will also evaluate high-flow conditions. The model should be able to simulate the range of conditions occurring under low flows and under high flow conditions.

•   The LA River modeling approach may be expanded in the future for TMDLs in downstream San Pedro Bay. Therefore, the chosen model should be capable of simulating estuaries or should allow for linkage or incorporation of another appropriate model or approach for addressing tidal systems.

2.2   Model Selection

Based on review of the guiding principles and the waterbody and nutrient related impairments, a list of selection criteria were identified the LA River application. The selection criteria define the specific model characteristics required to address the parameters set forth in the guiding assumptions and local conditions. As shown in Table 4, the selected model or series of models should be capable of simulating the hydrology and the water quality of the river system and should be capable of addressing the influential characteristics or aspects of the waterbody system. The model capabilities should be relevant to water quality issues of concern (e.g., algal growth) and the watershed and waterbody characteristics (e.g., nonpoint and point source inputs, low flows, etc.). Due to the many flow related factors that influence in-stream nutrient, algal, and dissolved oxygen concentrations, a hydrodynamic model linked with a water quality model is also needed.

Table 4. Criteria for Model Selection for the LA River Model Application

Model Type	Characteristics/Capabilities
Hydrodynamic	• Simulation of hydrology in tributaries and mainstem • Low flow or constant flow conditions • Variable flow (future applications) • Physical channel features (dams, weirs) • Incorporate point source inputs at specific locations • Ability to link to water quality model
Water Quality	• Nutrient cycle • Eutrophication processes • Algal growth • Benthic algae • Low flow or constant flow conditions • Variable flow (future applications) • Incorporate point source inputs at specific locations • Capability to simulate fecal coliforms and metals (future applications) • Ability to link to watershed loading models (future applications)

The modeling criteria and types of models were evaluated against the available models and recent applications of models for TMDL development. Model selection also considered access to models, model distribution and support, and acceptance by

EPA in similar TMDL applications. Based on the review a suite of models requiring minimal modifications were selected for the LA River application.

The 1-dimensional version of the hydrodynamic model Environmental Fluid Dynamics Code (EFDC) linked with the Water Quality Analysis Simulation Program (WASP) water quality model were selected for the LA River application. These models, both in the public domain and with a track record of TMDL applications, met most of the identified model selection criteria. The WASP model was modified slightly to meet the criteria for simulation of individual point sources and benthic algae. The following sections describe in more detail the models chosen for application in the LA River system, including why the models are the most appropriate for the analysis. Supplemental monitoring needs for application of the selected models were identified as well.

2.2.1 Hydrodynamic Model — EFDC

EFDC is a general purpose modeling package for simulating 1-D, 2-D, and 3-D flow and transport in surface water systems including rivers, lakes, estuaries, reservoirs, wetlands, and near shore to shelf-scale coastal regions. The EFDC model was originally developed at the Virginia Institute of Marine Science for estuarine and coastal applications, has been extensively tested and documented, and is considered public domain software.

In EFDC, a 1-dimensional variable cross-section sub-model solves the 1-D continuity, momentum, and transport equations within a variable cross-section framework. The 1-D sub-model uses the efficient numerical solution routines within the more general 2-D/3-D EFDC hydrodynamic model as well as the transport and meteorological forcing functions. Specific details on the model equations, solution techniques and assumptions can be found in Hamrick (1996).

The 1-D version of EFDC was used to simulate hydrodynamics in LA River and its tributaries. The 1-D version of EFDC was appropriate for use in the LA River analysis (as opposed to the 2-D or 3-D) because the evaluation focused on longitudinal changes in water quality conditions and data were not available to support use of the 2-D or 3-D versions of the model. The nature of the 1-D EFDC model as an extension of the more general 2-D/3-D model also provides the potential for direct linkage to future applications in the receiving waters at the confluence of the LA River with San Pedro Bay.

The use of variable cross-sections in EFDC makes it possible to use data available for the LA River channels to better define the channel and provide finer distinctions among channel segments, including areas of unlined channel and concrete channels. Because of the variable cross-section features, EFDC has the ability to account for the spreading grounds and the low flow channels in the LA River system. The ability to incorporate the spreading grounds in the system is important for the application of the model to future TMDLs considering wet weather conditions.

2.2.2 Water Quality Model — WASP5

EPA's Water Quality Analysis Simulation Program (WASP5) is an enhancement of the original WASP model (Di Toro et al., 1983; Connolly and Winfield, 1984; Ambrose, R.B. et al., 1988), which is a dynamic compartment model program for assessing aquatic systems, including both the water column and the underlying benthos. The time-varying processes of advection, dispersion, point and diffuse mass loading, and boundary exchange are represented in the basic program. Water quality processes are represented in special kinetic subroutines that are either chosen from a library or written by the user. WASP is structured to permit easy substitution of kinetic subroutines into the overall package to form problem-specific models. WASP5 permits the modeler to structure one, two, and three-dimensional models, allows the specification of time-variable exchange coefficients, advective flows, waste loads and water quality boundary conditions, and permits tailored structuring of the kinetic processes, all within the larger modeling framework without having to write or rewrite large sections of computer code.

EUTRO5 is a submodel of WASP5 that simulates the transport and transformation reactions of up to eight state variables related to eutrophication. They can be considered as four interacting systems: phytoplankton kinetics, the phosphorus cycle, the nitrogen cycle and the dissolved oxygen balance. The general WASP mass balance equation is solved for each state variable. To this general equation, the EUTRO5 subroutines add specific transformation processes to customize the WASP transport equation for the eight state variables in the water column and benthos. The eight state variables for EUTRO5 are:

- Nitrate/Nitrite
- Ammonia
- Organic Nitrogen
- Organic Phosphorus
- Orthophosphate
- Carbonaceous Oxygen Demand
- Phytoplankton (attached algae)
- Dissolved Oxygen

Figure 4 presents a schematic of the water quality kinetic processes simulated with the WASP Eutrophication model.

WASP was chosen for use in the modeling analysis of LA River because it can simulate all of the parameters of concern and it is easily linked with EFDC output. The linkage of EFDC and WASP permitted representation of major processes associated with nutrient cycling, algal uptake, and dissolved oxygen variability, including:

- Algal production through nutrient cycling and meteorological input
- Nitrification processes
- Oxygen production through photosynthesis
- Reaeration

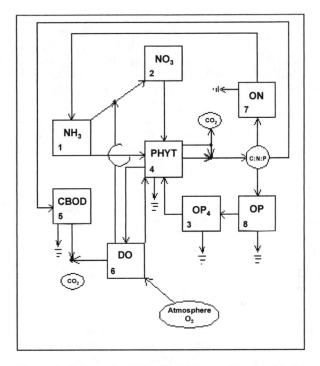

Figure 4. Schematic of WASP 5.1 Eutrophication Model

Several physical-chemical processes can affect the transport and interaction among nutrients, phytoplankton, carbonaceous material, and dissolved oxygen in the aquatic environment. EUTRO5 can be operated at various levels of complexity to simulate some or all of these variables and interactions. To simulate only BOD and DO, for example, the user may bypass calculations for the nitrogen, phosphorus, and phytoplankton variables. The following six levels of complexity are available for application:

(1) Streeter-Phelps
(2) Modified Streeter-Phelps
(3) Full linear DO balance
(4) Simple eutrophication kinetics
(5) Intermediate eutrophication kinetics
(6) Intermediate eutrophication kinetics with benthos

The first three levels of complexity only deal with the dissolved oxygen balance in the system. The "simple eutrophication kinetics" level is used to simulate the growth and death of phytoplankton interacting with only one of the nutrient cycles (i.e., nitrogen or phosphorous), while the "intermediate eutrophication kinetics" level simulates the growth and death of phytoplankton interacting with both the nitrogen and phosphorous cycles as well as the dissolved oxygen balance. Level six, "intermediate eutrophication kinetics with benthos," adds benthic interactions to the model simulations. To evaluate the effects of algae on the LA River system and both nitrogen and phosphorus, WASP was set up for intermediate eutrophication kinetics (level five). The effects of benthos were not assessed because the system is primarily lined with concrete providing limited habitat for benthic organisms.

2.2.3   Modifications to WASP

To accurately address the unique conditions in the LA River and its listed tributaries, the original WASP5 computer code was modified to allow for the following:

- Input of more than one load into a single segment
- Simulation of attached algae

The original WASP code limits the user to input only one load into any one segment. To input more than one load into a segment, these loads would be added together and the single combined load would have been used as input into the model. For most modeling applications this is sufficient. However, for the LA River and its listed tributaries, WASP was modified to input the loads separately, providing an efficient way to clearly identify and track each load input into the model.

Within EUTRO5 phytoplankton is modeled using chlorophyll-a as the input. Consequently, the subroutine for phytoplankton considers both movement in the water column (vertical) as well as movement between segments (horizontal). However, for the LA River, it was necessary to model phytoplankton that had no movement either vertically or horizontally (i.e., attached algae). The WASP subroutine for phytoplankton was modified to model attached algae, following the framework used by Warwick et al. (1997).

Mathematical relationships based on the impacts of temperature, available light, available nutrients, stream velocity, and density-dependant interactions were incorporated into the algae growth kinetics framework within EUTRO5. The major differences between modeling techniques for attached and free-floating algae are: (1) periphyton are expressed in terms of aerial densities rather than volumetric concentrations; (2) periphyton growth can be limited by the availability of bottom substrate; (3) the availability of nutrients to the periphyton matrix is influenced by current velocity; and (4) periphyton are not subject to transport.

The growth rate of attached periphyton was computed from the equation:

$$\frac{dP}{dt} = (AGR - R(t) - K1DP - K1GP \times INVT)P$$

where:
AGR	=	adjusted growth rate (d-1)
	=	G(T) x L(I) x min(L(DIN), L(DIP), L(V)) x L(D)
G(T)	=	maximum growth rate for temperature T
L(I)	=	light limitation factor

L(DIN)	=	dissolved inorganic nitrogen limitation factor
L(DIP)	=	dissolved inorganic phosphorus limitation factor
L(V)	=	velocity limitation factor
L(D)	=	density dependant growth factor reduction factor
P	=	periphyton biomass
R(T)	=	respiration rate for temperature T
K1DP	=	periphyton death rate
K1GP	=	grazing rate of periphyton
INVT	=	herbivorous invertebrate population grazing

The parameters used for the simulation of attached algae are presented in Table 5 in Section 3.1.4. The LA River simulations did not consider the effects of grazing on the growth of algae; therefore, the densities become a function of the growth rate balanced by the death rate and respiration.

Table 5. Parameters Used in the Calibration of the Water Quality Model

Name	Description	Value	Source
K12C	Nitrification (d^{-1})	1.00	Based on model calibration[1]
K12T	Nitrification temperature correction	1.080	Ambrose et al. 1991
KNIT	Nitrification oxygen limitation	2.00	Ambrose et al. 1991
K20C	Denitrification (d^{-1})	0.40	NDEP 208 Study
K20T	Denitrification temperature correction	1.045	Ambrose et al. 1991
PCRB	Phosphorous/carbon in periphyton	0.025	Caupp et al. 1991
NCRB	Nitrogen/carbon in periphyton	0.18	Caupp et al. 1991
KMPHY	Periphyton C half saturation (mg C/L)	0.005	Caupp et al. 1991
K1CP	Maximum periphyton growth (d^{-1})	0.851	Caupp et al. 1991
K1TP	Periphyton growth temperature correction	1.055	Caupp et al. 1991
KMNG1P	Periphyton N half saturation (mg N/L)	0.025	Caupp et al. 1991
KMPG1P	Periphyton P half saturation (mg P/L)	0.005	Caupp et al. 1991
KMVG1P	Periphyton velocity half saturation (m/s)	0.25	Caupp et al. 1991
KBP	Periphyton density half saturation (g C/m^2)	6.5	Warwick et al. 1997
K1RCP	Periphyton respiration (d^{-1})	0.0175	Jorgensen 1979
K1RTP	Periphyton respiration temperature correction	0.0690	Jorgensen 1979
K1DP	Periphyton death (d^{-1})	0.02	Ambrose et al. 1991

Table 5. Parameters Used in the Calibration of the Water Quality Model (con't)

KVMINP	Periphyton velocity limitation minimum (m/s)	0.15	Whitford et al. 1964
OCRB	Oxygen/Carbon in periphyton	2.67	Ambrose et al. 1991
K71C	Organic-N mineralization (d^{-1})	0.50	Warwick et al. 1997
K71T	Organic-N minimum temperature correction	1.080	Ambrose et al. 1991
FON	Fraction of nonrecycled Org-N	0.15	Warwick et al. 1997
K83C	Organic-P mineralization (d^{-1})	0.75	Warwick et al. 1997
K83T	Organic-P minimum temperature correction	1.080	Ambrose et al. 1991
FOP	Fraction of nonrecycled Org-P	0.50	Ambrose et al. 1991

[1]A sensitivity analysis was conducted to evaluate the effects of nitrification values on the model calibration. Further discussion is included in Section 3.4.

2.3   Supplemental Monitoring

This TMDL focuses on the critical period for nutrient loading to the LA River system—times of low flow when point sources provide the majority of the instream flow. During low flow conditions, the three major WWRPs comprise 60 to 80 percent of the river's flow and approximately 80 percent of the nitrogen loading. In addition to the major WWRPs, other dry-weather sources that deliver flow and nutrients to the LA River system include storm drain discharges (e.g., dry weather runoff from residential and commercial water use), tributaries, and groundwater.

To evaluate the loading, transport, and resulting water quality effects of nitrogen compounds in the LA River system it is necessary to characterize and account for each of the sources in the model. However, data were not available to appropriately characterize all of the sources. Previously collected data focused on larger segments of the mainstem and on the major point sources. Flow and water quality data were readily available for the point sources and were used for model input for the WWRPs. However, available data did not provide the information necessary to define the smaller inputs to the system. To better characterize the sources influencing flow and water quality in the LA River system, SCCWRP conducted intensive monitoring events in the watershed in September 2000 and July 2001 during periods representative of typical low-flow conditions. The first monitoring event was conducted on September 10 and 11, 2000, and the second was conducted on July 29 and 30, 2001. The datasets collected represent snapshots of the flow distribution and water quality conditions throughout the LA River system.

Data collected by SCCWRP included measurements of flow and water quality at the following locations:

- LA County Department of Public Works (LACDPW) monitoring stations
- Mainstem and tributary headwaters
- Confluence of tributaries and the mainstem LA River (2001 only)
- Dry-weather stormwater inputs on the mainstem LA River (33 locations in 2000; 69 locations in 2001)
- Dry-weather stormwater inputs on the 303(d)-listed tributaries (15 locations in 2000; no locations in 2001)

Data were collected by SCCWRP for use as model input as well as for comparison to model results during calibration and validation. Flow and water quality measurements collected at tributary stations (headwater or confluence stations) during the intensive monitoring efforts were used as model input to represent the tributary discharges into the LA River mainstem. Flow and water quality data were also collected for identified dry-weather stormwater flows during the September 2000 and July 2001 monitoring and used as model input to represent the dry-weather discharges from stormdrains.

In addition to data used as input to the model, SCCWRP collected data to provide instream flow and water quality measurements to compare with model results during model calibration and validation. During the summer of 2000, SCCWRP performed three dye studies in the LA River mainstem to collect data to use in calibrating the hydrodynamic model for velocity. Flow and water quality measurements collected at LACDPW mainstem stations during September 2000 were used for model calibration while data collected during the July 2001 monitoring efforts were used for model validation. For all of the stations, triplicate composite samples were collected at each location to provide a measure of the system variability for water quality calibration. The following parameters were analyzed to cover the full list of state variables simulated in the WASP water quality model:

- ammonia nitrogen
- nitrate nitrogen
- organic nitrogen
- orthophosphate
- organic phosphorus
- dissolved organic carbon
- particulate organic carbon
- chlorophyll-a (as an attached algal component)
- dissolved oxygen

The use of the data for model calibration and validation is discussed in the following sections.

3. Model Development for the Los Angeles River

The selected models were applied to the LA River system according to a standard modeling strategy. The following steps were executed in the development of the model.

- Model configuration and identification of application conditions, including model linkages, simulation period, model boundaries and all model inputs.
- Sensitivity analysis.
- Hydrodynamic calibration and validation.
- Water quality calibration and validation.

These sequential steps are designed to build the modeling system and provide testing and evaluation of model performance at each step. The initial configuration stage defines the essential structure of the modeling network. This is where the river is divided into "segments," or units for analysis, and the locations of all the various inputs are defined. During this step the various input data are compiled and the time periods for analysis are defined. Next the hydrodynamic portion of the analysis is performed. The hydrodynamic application is first "calibrated," by using the best available information and adjusting parameters within reasonable range to achieve the best fit with the observed data. Next, the hydrodynamic application is "validated" by testing the input file (without adjustment) with another time period. Once the hydrodynamic validation is complete, the calibration/validation process is performed for the water quality simulation. A supplemental sensitivity analysis can also be used to explore or evaluate the response of the model to changes in selected parameters. The sensitivity analysis can be used to test and evaluate options in the setup or configuration of the model. This sequence of application and testing is used to build a modeling system that is representative of local conditions and able to evaluate the various management scenarios.

The following sections describe each of the key steps in the model development process.

3.1   Model Configuration and Application Conditions

The following subsections describe the model set-up for the LA River system, including model linkages, simulation period, model boundaries, and model input parameters.

3.1.1   Model Linkages

The 1-D EFDC model was utilized to simulate the flow and transport within the LA River under dry weather conditions. The nutrient cycling and algal growth processes were simulated using the EUTRO5 component of the WASP5 model system. The EFDC model was externally linked to the WASP model through a hydrodynamic forcing file that contains the flows, volumes, and exchange coefficients between adjacent cells. The EFDC model takes the user-defined flow inputs (e.g., point source discharges, dry-weather stormdrain discharges, etc.) and develops in-stream flows and transport that are passed to the WASP5 model through a hydrodynamic linkage file. The WASP water quality model then runs on a similar time step with the same grid network layout. Figure 5 presents a schematic of the instream model network used throughout this study, with the reaches shown corresponding to the listed segments within the LA River watershed.

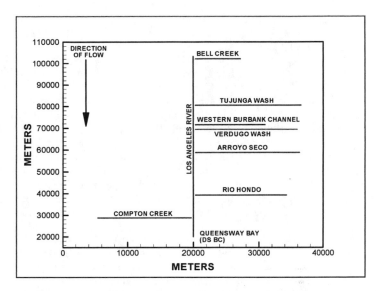

Figure 5. Schematic Representation of EFDC 1-D Model Grid

3.1.2 Simulation Period

Selection of simulation periods for the LA River TMDLs was dependent on the critical conditions for nitrogen compounds and related effects, which are low flow periods (April to September) when point sources dominate the instream flow and nutrient loading. Because data were limited to characterize tributary and dry-weather inflows under critical conditions, SCCWRP conducted two intensive monitoring efforts throughout the watershed to better understand these "unmeasured" inflows.

Simulation periods for model calibration and validation correspond to the dates of the two monitoring efforts. Model calibration was performed for September 10 and 11, 2000, and model validation was performed for July 29 and 30, 2001. The TMDL scenarios were evaluated under the calibration conditions of September 10-11, 2000.

3.1.3 Downstream Boundary

The downstream boundary used for the hydrodynamic simulations was the tidal signal from the Long Beach Inner Harbor Tide Station. The tidal signal in the LA River does not impact the areas of concern for this study, but the boundary was set with the intention of providing future links to hydrodynamic and water quality models in the harbor area.

3.1.4 Model Setup and Inputs

The following describes data that were used in the model setup and the inputs used in the 1997, 2000 and 2001 simulations for low flow conditions. These include the following hydrodynamic (EFDC) and water quality (WASP5) inputs:

- Geometry
- Topography
- Meterological data
- Source data

3.1.4.1 Geometry

All of the waterways modeled were concrete lined except for a small segment of the LA River near Glendale where a high groundwater table prevents the placement of concrete and the area of the Sepulveda Basin. The major waterways in the LA River watershed were planned and constructed in the early part of the twentieth century. Over time, modifications have been made to the LA River watershed conduit system such as adding low flow channel sections, repairing deteriorated portions, and other various as-needed work. As a result of the size of the watershed conduit system and time period for the majority of the construction, there was not a readily discernible location for complete and current geometric information on the major waterways.

However, detailed geometry data were needed to physically define the LA River system in the models to appropriately simulate flow and transport under low-flow conditions. The model of the LA River and tributaries was established with a variable cross-section grid and a total of 302 grid cells averaging 600 meters in length. For these cross-sections geometric input files were established for the model with the following user-defined information:

- Invert elevation
- A range of depths measured above the invert, covering the full depth of the cross-section
- Cross-sectional area associated with each depth above the invert
- Wetted perimeter associated with each depth above the invert
- Top width associated with each depth above the invert

The geometric input files represent the full cross-section of the river, including the low-flow channel. The EFDC model is then capable of simulating the full range of flow conditions that do not overtop the existing channel.

Invert elevation and cross sectional geometry for the waterways in this study were determined from review of approximately 1,500 construction plans and as-built drawings, approximately 80 typical section sheets from the LACDA USACE O&M Manual, approximately 20 FEMA flood study HEC-2 decks, photographs, and limited field reconnaissance.

Figures 6 and 7 show photographs of two sections along the main stem LA River and provide examples of cross-sectional variation throughout the system. As the photos show, the channel geometry changes significantly throughout the system. For the model segmentation the cross-sections remained constant until alternate sections were defined within the as-built drawings or other sources.

Figure 6. Channel Cross-Section at LA River Station 2

Figure 7. Channel Cross-Section at LA River Station 6

Within the main stem LA River and listed tributaries, grids were established to correspond to areas of changing cross-section, slope or channel characteristics.

3.1.4.2   Topography

The topography of the LA River watershed is represented by two distinct areas, the very steep mountain regions in the Santa Monica, Santa Susana and San Gabriel Mountains, and the low lying relatively flat sections in the San Fernando Valley and the lower LA River. Topographic data used in the model simulations were obtained from the USGS Digital Elevation Model (DEM) within the BASINS database with a resolution of 90 x 90 meters. Figure 8 presents the DEM data used in the model simulations. Elevations within the watershed range from near sea level at the lower reaches of the LA River to greater than 2,000 meters above sea level. Within the LA River model network, the DEM provided invert elevations and slopes for the channel sections where data were not available from the as built drawings.

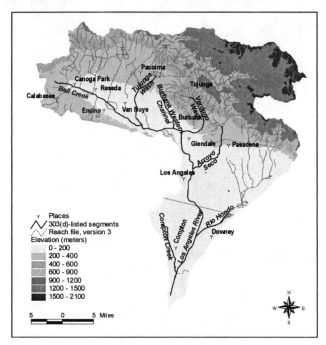

Figure 8. Topography in the LA River Watershed

3.1.4.3   Meteorological Data

Relevant meteorological parameters necessary for input into EFDC and WASP models are:

- Air Temperature
- Relative Humidity
- Wind Speed
- Wind Direction
- Solar Radiation
- Cloud Cover

The primary weather station located at the Los Angeles Airport, provided the meteorological data used in the simulation of temperature in the EFDC hydrodynamic model and algal growth in the WASP water quality model. Given the nature of this type of data, a single station located at the airport was sufficient because spatial variability is not as critical for these parameters as it is to rainfall. Because the modeling evaluates dry-weather conditions with no rain-driven inputs, precipitation data are not a necessary input for the LA River low-flow modeling. However, all meteorological data were input to the models for completeness.

3.1.4.4   Water Quality Parameters

This section presents the water quality model parameters used in the WASP model simulations. These values are included in Table 5 and represent default values available for WASP, based on accepted literature values. Table 5 includes parameters that were used to simulate attached algae in the LA River. The attached algae subroutines that were added to the WASP code followed the methodology presented by Warwick et al. (1997). During the September 2000 and July 2001 monitoring, SCCWRP measured algal biomass and nutrient uptake at four locations along the LA River and at the confluence of Arroyo Seco. However, this limited data were inconclusive about the algae in the LA River. Therefore, the parameters from Warwick (1997) were used in the calibration.

3.1.4.5   Source Representation

The setup of the modeling system also requires the initial representation of the various sources of flow and constituent loading to the system for the simulation time periods. This initial representation of the sources is based on a combination of historic monitoring and information gathering, new targeted data collection, and mass balance analysis. For this application SCCWRP conducted targeted monitoring throughout the LA River watershed in September 2000 and July 2001 to better characterize sources of flow and nutrients to the LA River. This section discusses the supplemental data gathering, the analysis of the available data, and how this information was used to best represent sources in the models.

Examination of the LA River system indicates that the following potential sources and sinks of flow and constituent loading are present:

- Point Source Discharges
- Stormwater Inflows
- Tributary Inflows
- Groundwater (recharge or infiltration)

The analysis of historic data was used to determine when various sources are active and the potential distribution of flow contributions. Examination of instream flow data from LACDPW and the City of Los Angeles was used to determine the flow distribution and patterns in the LA River system. Eleven stations had data available during the 1997 and 2000 water years (Figure 9). Of these 11 stations, all but one (F319-R) had data available at 15-minute intervals. Station F319-R had daily average data.

Figure 9. Flow Measurement Stations in the LA River and its Tributaries

Examination of historic flow records and point source discharge records confirm that a significant source of flows during low flow periods are point source discharges. Presently there are six major permitted point source discharges to the LA River and its tributaries, and 29 minor permitted discharges. Table 6 presents a list of the major and minor dischargers along with their NPDES permit numbers and design flows and Figure 10 presents the locations of the major discharges.

Table 6. NPDES Permitted Major and Minor Discharges (LARWQCB, 2000)

NPDES#	Discharger	Facility	Design Q (mgd)	Class
CA0001309	The Boeing Company	Rocketdyne Div. - Santa Susana	15.000000	MAJOR
CA0052949	Southern California Edison	Dominguez Hills Fuel Oil Fac	4.320000	MAJOR
CA0053953	LA City Bureau of Sanitation	L.A.-Glendale WWRP, NPDES	20.000000	MAJOR
CA0055531	Burbank, City Of Public Works	Burbank WWRP, NPDES	9.000000	MAJOR

Table 6. NPDES Permitted Major and Minor Discharges (LARWQCB, 2000) (con't)				
CA0056227	LA City Bureau of Sanitation	Tillman WWRP, NPDES	80.000000	MAJOR
CA0064271	Las Virgenes MWD	Tapia Park WWRP, NPDES	2.000000	MAJOR
CA0000892	Kaiser Aluminum Extruded Prod.	Kaiser Aluminum Extruded Prod.	0.125000	MINOR
CA0001899	Celotex Corporation	Asphalt Roofing Mfg, La	0.120000	MINOR
CA0002739	MCA / Universal City Studios	Universal City Studios	0.169000	MINOR
CA0003344	Kaiser Marquardt, Inc.	Ramjet Testing, Van Nuys	0.024000	MINOR
CA0056464	Owens-Brockway Glass Container	Glass Container Div, Vernon	0.408100	MINOR
CA0056545	Los Angeles City Of Rec&Parks	Los Angeles Zoo Griffith Park	2.010000	MINOR
CA0056855	Los Angeles City of DWP	General Office Building	1.500000	MINOR
CA0057274	Pabco Paper Products	Paperboard & Carton Mfg,Vernon	0.745800	MINOR
CA0057363	Edington Oil Co.	Long Beach Refinery - Rainfall	0.560000	MINOR
CA0057690	Bank Of America	Nt & Sa L.A. Data Center	0.015000	MINOR
CA0057886	Filtrol Corp.	Filtrol Corp.	0.897000	MINOR
CA0058971	Exxon Co., U.S.A.	Exxon Company U.S.A.	0.032000	MINOR
CA0059242	Consolidated Drum Recondition	Oil Drum Recycling, South Gate	0.008500	MINOR
CA0059293	Chevron U.S.A. Inc.	Van Nuys Terminal	0.050000	MINOR
CA0059561	Arco Terminal Services Corp.	East Hynes Tank Farm	0.190000	MINOR
CA0059633	Metropolitan Water Dist. Of SC	Rio Hondo Power Plant	0.050000	MINOR
CA0062022	Dial Corp, The	The Dial Corporation	0.028800	MINOR
CA0063312	3M Pharmaceuticals	3M Pharmaceuticals	0.144000	MINOR
CA0063355	Pasadena, City Of, DWP	Dept. Of Water & Power	0.411000	MINOR

Table 6. NPDES Permitted Major and Minor Discharges (LARWQCB, 2000) (con't)

CA0063908	McWhorter Technologies, Inc.	McWhorter Technologies, Inc.	0.075000	MINOR
CA0064025	Sta - Lube, Inc.	Sta - Lube, Inc.	0.150000	MINOR
CA0064068	Lincoln Avenue Water Co.	South Coulter Water Treatment	0.018500	MINOR
CA0064084	Mairoll, Inc.	Voi-Shan Chatsworth	0.014400	MINOR
CA0064092	Los Angeles County MTA	Metro Lines-Segments 1 & 2a	0.500000	MINOR
CA0064149	Los Angeles City of DWP	Tunnel # 105	0.005900	MINOR
CA0064190	Pacific Refining Co.	Former Western Fuel Oil	0.001200	MINOR
CA0064203	Los Angeles Turf Club	Santa Anita Park	12.700000	MINOR
CA0064238	Water Replenishment Dist Of S.C	West Coast Basin Desalter	2.200000	MINOR
CA0064319	Coltec Industries Inc.	Former Menasco Aerosystem Faci	0.014000	MINOR

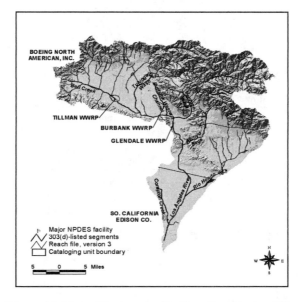

Figure 10. Major Wastewater Reclamation Plants within the LA River Watershed

If all of the major permitted facilities discharged at their design flow conditions, they would account for approximately 85 percent of the point source inputs to the LA River (Figure 11). Because many of the minors are stormwater-related, their contribution during dry periods is negligible. Additionally, examination of the design flows for the Glendale, Tillman, and Burbank WWRPs in relation to the other three majors shows that these three facilities account for over 80 percent of the major design discharge (Figure 11). Additionally, the Boeing and SC Edison discharges are primarily storm water and their contributions during dry weather are negligible. The Las Virgenes facility has a special permit that allows them to discharge to the LA River during high flow events. During the period used in the simulations, they did not exercise this option to discharge, and therefore the discharge from the Las Virgenes facility was not included in the model. Therefore the only point sources included in the low-flow simulations of the LA River system are the Glendale, Tillman, and Burbank WWRPs.

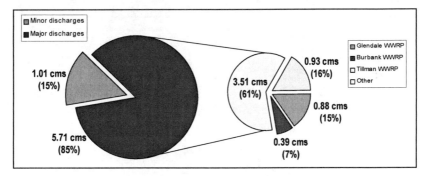

**Figure 11. Distribution of Design Flows Between the Major
and Minor Discharges**

Analysis of the data showed that during the dry periods, point source discharges accounted for 60 to 100 percent of the total flow through the system. The remaining flows were attributed to groundwater inflow, discharge from dams upstream from the listed segments, and residential, commercial and industrial water uses. The gauged tributary data account for some of the additional 20 to 40 percent of the dry weather base flow on the main stem LA River, but additional flow still remains unaccounted for based on these measurements.

To support the analysis of historic flow data, the in-stream model (EFDC) was also used to investigate potential sources of flow during the 1997 low-flow period. Evaluation of the mass balance of flow in the system during the 1997 model testing showed that the sources of up to 40 percent of the total flow in the system during low flow conditions were unknown. Because no data were available to quantify the additional flows to the system (e.g., dry-weather stormwater inputs) during 1997, assumptions were made about the quantity and distribution of inflows to achieve reasonable comparison with the measured flows at the bottom of the system. The model testing for 1997 was not intended to calibrate the model to observed values, but rather to provide qualified estimates of the flow distribution in the system and help establish additional data needs.

Figure 12 present comparisons of the measured versus simulated flows at four stations throughout the system (see Figure 9 for station locations) for the 1997 low flow period. The comparison stations represent four locations along the main stem of the LA River. The simulated and measured flows range from 75 to 100 cfs at the upper most station (F300-R) to between 125 and 150 cfs at the lowest station (F319-R). The lowest station (F319-R) is below the confluence of all tributaries within the LA River and all simulated point source discharges. This station reflects the total water "mass balance" within the system under the relatively steady low flow condition. This simulation was to provide a preliminary testing of the model to get results with similar patterns and magnitudes as observed data. Differences in flow are likely attributable to stormwater flows and other unknown flows that were not specifically included in this simulation.

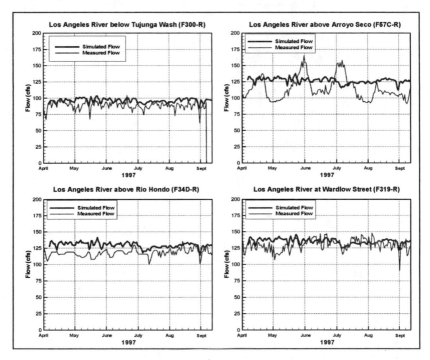

Figure 12. Simulated vs. Measured Dry Weather Flow, 1997 (EFDC)

Another important source of flow that could be a large portion of the "unknown" flow in the system is dry-weather inputs from stormdrains. Minor residential and commercial stormwater flows are typically a small portion of the total water budget during the wet weather period, but can be a considerable percentage during the dry weather period. The stormwater inputs during dry-weather periods represent inflows from the stormwater conveyances throughout the system from sources such as golf courses, car washes or residential lawns. Data collected by SCCWRP in September 2000 and July 2001 indicated that storm drain flows contributed 7 to 15 percent of the total nitrogen load to the LA River on the monitored days. This information illustrates the importance of capturing the inputs from storm drains in the models representing the LA River system. The 2000 and 2001 SCCWRP data were used to develop flow and water quality model inputs for dry weather stormwater discharges in the watershed.

At times, comparison of the flows from the three major point source discharges exceeded the total flow measured at the stations downstream of all of the inflow points (Stations F57C-R, F34D-R, and F319-R). During these time periods three possible explanations exist for the conditions:

•   Errors in the gauging stations in measuring very low flow conditions
•   Evaporative losses within the system
•   Losses due to groundwater recharge

It may be that during these time periods all three of these processes are occurring and the result is a net loss of water in the system.

The models were set-up to account for all of the potential sources of flow and nutrients in the LA River system. The following sections discuss the data used to represent the hydrodynamics and water quality of each of the major sources— WWRPs, tributaries, dry-weather stormwater and groundwater.

3.1.4.6   Hydrodynamic Data Used for Source Representation

The EFDC hydrodynamic model was calibrated and validated for application to the LA River system. The model was calibrated to observed data collected during the monitoring effort on September 10 and 11, 2000. After the model was calibrated, model validation was performed using the data set collected on July 29 and 30, 2001.

For each of these hydrodynamic simulations, it was necessary to characterize the sources of flow as closely as possible to the conditions occurring during the simulation period. Table 7 presents a summary of the representation of inflows in the models and the following sections provide more details on the data used as input for the model calibration and validation, including data used to characterize and represent inputs to the LA River from the main sources of flow¾WWRP, tributary, stormwater and groundwater flows. Section 3.3 presents and discusses the results of the testing, calibration and validation.

Table 7. Summary of hydrodynamic representation of sources in the
LA River system

Source Inflows	Representation in Calibration	Representation in Validation
WWRP	Constant flow based on measured daily average flow of WWRP effluent on September 10-11, 2000	Constant flow based on measured daily average flow of WWRP effluent on July 29-30, 2001
Tributaries	Constant flows based on flows measured at tributary headwaters on September 10-11, 2000	Constant flows based on flows measured at tributary confluences on July 29-30, 2001
Dry-weather Stormwater	Constant flows based on measured flows of 48 identified stormwater flows on September 10-11, 2000	Constant flows based on measured flows of 69 identified stormwater flows on July 29-30, 2001
Groundwater	Infiltration, based on mass balance evaluation	Not included, based on mass balance evaluation

WWRP Flow

Point source discharges provide a substantial portion of the LA River system's flow. Therefore it is necessary to include inputs in the model to represent discharges from the major point sources Glendale WWRP, Burbank WWRP, and Tillman WWRP. The following discusses how the WWRP discharges were represented in the EFDC hydrodynamic model for the model calibration and validation.

Flow data obtained from the three major WWRPs (Tillman, Glendale and Burbank) were used as input to EFDC for model calibration and validation. Daily average flows measured by the WWRPs for September 10 and 11, 2000, were used as constant flows representing their respective discharges in the EFDC model calibration and flow data from the WWRPs for July 29 and 30, 2001, were for model validation.

The Glendale WWRP had only one discharge location, the Burbank WWRP had two outlets, both discharging to Burbank Western Channel, and the Tillman WWRP effluent is discharged to the LA River through the following four outlets:

- Direct discharge to the LA River
- Discharge to the Wildlife Lake with eventual outflow to the LA River
- Discharge to the Recreation Lake within the Sepulveda Basin with eventual outflow to the LA River
- Discharge to the Japanese Tea Gardens, with eventual feedback to the direct discharge

Table 8 presents the WWRP flows used in the model for calibration and validation. These values are used as constant flow values in the model to represent the discharges from the WWRPs to the LA River system.

Table 8. Flow Data from the Three Major Point Source Discharges Used in Model Calibration and Validation

Point Source Discharge	Flows used in Calibration[1]		Flows used in Validation[1]	
	Flow (cms)	Flow (mgd)	Flow (cms)	Flow (mgd)
Tillman WWRP — Direct Discharge	1.507	34.4	0.407	9.3
Tillman WWRP — Japanese Gardens	0.210	4.8	0.197	4.5
Tillman WWRP — Recreation Lake	0.762	17.4	0.762	17.4
Tillman WWRP — Wildlife Lake	0.258	5.9	0.250	5.7
Tillman WWRP — TOTAL	2.737	62.5	1.617	36.9
Glendale WWRP	0.407	9.3	0.403	9.2
Burbank WWRP	0.403	9.2	0.269	6.2

[1]Based on discharge monitoring data provided by the WWRP

Tributary Inflows

In addition to flow contributions from the WWRPs, the LA River system receives flow from tributary inflows and baseflows during low flow periods. These flows arc included in EFDC with a representative constant flow value that was defined using monitoring data or evaluation of a mass balance when monitoring data were unavailable. Table 9 presents the values used to represent tributary inflows in the model setup, calibration and validation, and the following paragraphs provide further discussion on the identification of these flow values. All flows were input to the uppermost cell of each segment (e.g., mainstem, Compton Creek) as constant flows.

SCCWRP monitoring data were used to identify model inputs for the tributary inflows for the hydrodynamic model calibration and validation. SCCWRP included monitoring in the upper reaches of all the listed tributaries as part of the September 10-11, 2000, intensive monitoring. Figure 13 shows the locations of the headwater monitoring stations. Table 9 presents the measured headwater flow data for the LA River and its tributaries. These values were used in the model calibration to characterize the flow contributions from the tributaries as constant discharges.

**Table 9. Measured Tributary Inflows Used for Model
Calibration and Validation**

Location	Flows Used in Calibration[1] (cfs)	Flows Used in Validation[2] (cfs)
Main Stem LA River	10.59	Not included
Compton Creek	2.97	1.80
Rio Hondo	NO FLOW	NO FLOW
Arroyo Seco	0.00	3.32
Verdugo Wash	1.36	2.20
Burbank Channel	1.41	9.51
Tujunga Wash	0.67	0.37
Bell Creek	1.20	2.65

[1]Based on data collected at headwater stations by SCCWRP on September 11-12, 2000.
[2]Based on data collected at confluence stations by SCCWRP on July 29-30, 2001.

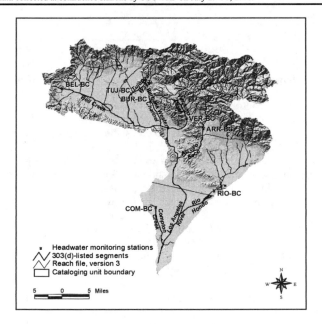

Figure 13. Headwater Flow Monitoring Stations

During the July 2001 data collection, SCCWRP measured flows at the upper reaches
of all of the listed tributaries as well as at their confluence with the LA River. The

locations of the headwater stations were the same as those used in the 2000 dataset (Figure 13). Because the model is steady-state, if there are no additional inflows to a tributary (e.g., stormwater inputs), the modeled flow at the bottom of the tributary (i.e., at the confluence) is equal to the input flow at the headwaters. To better represent the inflows from the tributaries, 2001 confluence flows were used to define the tributary flows for the model validation instead of the headwater flows. Table 9 presents the flows measured at the confluences of each of the tributaries and the LA River and used as constant flow inputs in the model to represent flow contributions from tributaries for model validation.

Stormwater Inflow

Another source of flow contributions to the LA River system are dry-weather stormwater flows. Minor stormwater flows are typically a small portion of the water budget during wet-weather periods but can be a considerable percentage during dry weather periods. During the 2000 data collection by SCCWRP, 67 dry-weather stormwater flows were identified in the LA River and its tributaries (Figure 14). Flow was measured at 48 of the 67 total flows identified and input into the hydrodynamic model to represent flow contributions from dry-weather stormwater during the calibration period. The remaining identified stormwater flows represent locations where the flows could not be measured (e.g., the flows were too small or had already moved downstream) and were therefore not included as inputs to the model. Figure 14 presents the locations of the dry-weather stormwater inflows during the September 2000 sampling event with a summary of their spatial distribution. Table 10 presents a summary of the totals of the individual dry-weather stormwater flows included in the model by listed reach.

Table 10. Totals of Measured Dry-Weather Stormwater Inflows to the Listed Reaches in the LA River Watershed

Location	Total of Individual Dry-Weather Stormwater Flows (cfs)	
	Used in Calibration[1]	Used in Validation[2]
Main Stem LA River	25.58	64.19
Compton Creek	0.10	Not Measured
Rio Hondo	NO FLOW	Not Measured
Arroyo Seco	3.72	Not Measured
Verdugo Wash	1.46	Not Measured
Burbank Channel	0.00	Not Measured
Tujunga Wash	0.00	Not Measured
Bell Creek	3.09	Not Measured

[1]Based on data collected by SCCWRP on September 11-12, 2000.
[2]Based on data collected by SCCWRP on July 29-30, 2001.

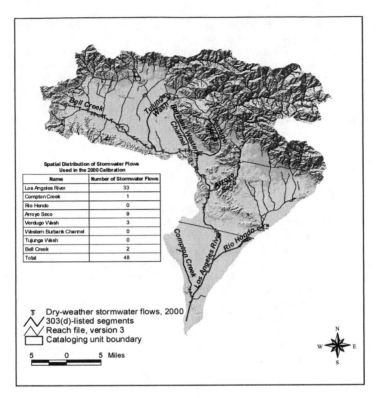

Figure 14. Stormwater Inflow Measurements during September 10-11, 2000

During the July 29-30, 2001, monitoring effort, SCCWRP again measured dry-weather stormwater flows to be included in the models, this time for model validation (Figure 15). Unlike the September 2000 data collection, dry-weather stormwater inflows were collected only on the LA River and not on the tributaries. Because September 2000 data suggested that dry-weather stormwater inflows on the tributaries were insignificant during the low flow period, more effort was spent on quantifying dry-weather stormwater inflows on the LA River. During the 2001 data collection, 105 dry-weather stormwater flows were identified in the LA River. Flow was measured at 69 of the 105 total flows identified and input in the model. The remaining 36 flows represent flows that could not be measured and are not included in the model. Table 10 provides a summary of the stormwater flows represented in the model, presenting the total of the individual flows into each listed reach.

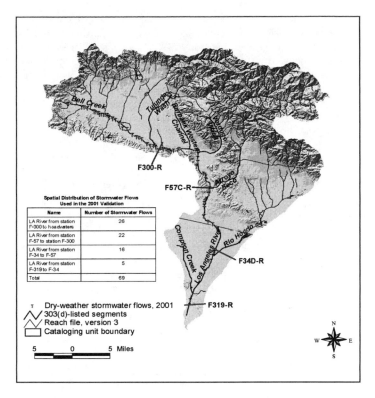

Figure 15. Stormwater Inflow Measurements during July 29-30, 2001

Groundwater

Groundwater recharge can add water to the system while infiltration can cause a flow loss. While data are not available to directly measure the groundwater component of the LA River system, the net groundwater contribution can be estimated using a mass balance of known flows in the system. Although it is likely that groundwater is a small portion of the flow budget in the system, the mass balance was used to estimate its magnitude and it was assumed that the "leftover" flow input or loss necessary to achieve mass balance is the groundwater component in the system. Mass balances were performed for the time periods of the model calibration and validation to identify the gain or loss attributed to groundwater and to account for this flow component in the model. Adjustments included the following:

- For the low flow simulations for the model calibration the base flows within the LA River and the tributaries were based on the flow measurements at the upstream end of the tributaries and the measurement of intermittent stormwater flows coming into the system on September 10-11, 2001. The groundwater interaction at Glendale Narrows was assumed to be a net decrease of water (infiltration) into the unlined portion of the river at the Narrows. The total flow at station F57-R (Los Angeles River at Arroyo Seco), which is located just below the Glendale Narrows, was approximately 0.52 cms less than the measured data. This difference was input into the hydrodynamic model as infiltration for the model calibration.

•   During the 2001 validation period, the total flow measured at station F-319 (Los Angeles River at Wardlow Rd) showed that during the two days of data collection, the sum of the measured flows fell within the range measured at station F-319. This indicates that a mass balance was achieved, and infiltration or recharge was not input into the model for the validation period.

3.1.4.7   Water Quality Data Used for Source Representation

The WASP model was calibrated to the conditions measured on September 10 and 11, 2000, and then validated to the conditions measured on July 29 and 30, 2001, corresponding to the dates of the SCCWRP monitoring efforts. Table 11 presents a summary of the model representation of water quality inputs from sources of nutrients to the LA River system - WWRPs, tributaries, and stormwater concentrations - and this section presents further discussion of the data used to characterize the inputs for the WASP water quality calibration and validation.

Table 11. Summary of Water Quality Representation of Sources in the LA River System

Source Inflows	Representation in Calibration	Representation in Validation
WWRP	Nutrient concentrations based on measurements of WWRP effluent on September 10-11, 2000	Nutrient concentrations based on measurements of WWRP effluent on July 29-30, 2001
Tributaries	Nutrient concentrations based on water quality measurements at tributary headwaters on September 10-11, 2000	Nutrient concentrations based on water quality measurements at tributary confluences on July 29-30, 2001
Dry-weather stormwater	Nutrient concentrations based on water quality measurements in 48 identified stormwater flows on September 10-11, 2000	Nutrient concentrations based on water quality measurements in 69 identified stormwater flows on July 29-30, 2001
Groundwater	Infiltration, based on mass balance evaluation	Not included, based on mass balance evaluation

WWRP Water Quality

The Tillman, Glendale and Burbank WWRPs represent a significant portion of the flow and nutrient contributions to the LA River system. The WWRPs in the LA River watershed routinely monitor their discharge effluent. Water quality data for the three major WWRPs collected on September 11, 2000, were used as input into the WASP model for calibration and effluent measurements from July 30, 2001, were used as input for the model validation. Table 12 presents the water quality data used to represent WWRP discharges in the model calibration and validation. Because dissolved oxygen data were not collected at the point source discharges on the days corresponding to the calibration and validation simulation periods, 6.0 mg/L was used as input to the model. (A sensitivity analysis was performed on the model using dissolved oxygen values from 2.0 mg/L to 10 mg/L, resulting in minimal effect on model results.)

Table 12. Water Quality Characteristics of WWRP Inputs for Model Calibration and Validation

Point Source Discharge		Ammonia (mg/L)		Nitrate+ Nitrite (mg/L)		Organic Nitrogen (mg/L)		Ortho-phosphate (mg/L)		Organic Phos-phorous (mg/L)	
		Cal.	Val.	Cal.	Val.	Cal.	Val.	Cal.	Val.	Cal.	Val.
Tillman WWRP	Direct Discharge	13.40	15.37	0.10	1.39	1.80	2.26	1.56	1.90	0.15	0.18
	Japanese Gardens	12.50	16.26	0.90	0.77	3.10	2.71	1.59	2.36	0.15	0.22
	Recreation Lake	4.35	5.50	7.55	5.75	4.30	4.37	0.96	2.07	0.15	0.32
	Wildlife Lake	12.50	16.26	0.90	0.77	3.10	2.71	1.59	2.36	0.15	0.22
Glendale WWRP		3.67	10.30	2.69	3.65	1.00	2.63	1.62	2.65	0.01	0.01
Burbank WWRP		19.00	18.35	0.50	0.52	2.00	1.42	0.50	1.27	0.50	1.27

The Tillman plant does not discharge directly to the LA River, but first passes through three other discharges: Japanese Gardens, Recreation Lake and Wildlife Lake. The four discharges were included in the model with individual characteristics (Table 12).

Tributary Water Quality

Water quality data collected by SCCWRP at the upstream boundaries of the listed tributaries were used to define nutrient inputs from tributaries for model calibration and validation. Data were collected on September 10-11, 2000, and July 29-30, 2001, at the headwater stations shown in Figure 16. The data at each boundary consisted of three composite samples. Three grab samples were taken to create each composite sample. A third of each grab sample was then combined into one bottle forming the composite sample. The purpose of this method for collecting water quality data was to eliminate the variability that occurs in sampling, as well as the variability that occurs in the river. Table 13 presents the water quality data used to represent tributary inflows in the model calibration and validation. The data were input into the model as nutrient concentrations at the upstream boundary of each listed tributary.

Figure 16. Location of Tributary and Instream Water Quality Measurements

Table 13. Water Quality Concentrations of Inflows from Tributaries for Model Calibration and Validation

Tributary	Ammonia (mg/L)		Nitrate+ Nitrite (mg/L)		Organic Nitrogen (mg/L)		Ortho-phosphate (mg/L)		Organic Phosphorus (mg/L)	
	Cal.[1]	Val.[2]	Cal.[1]	Val.[2]	Cal.[1]	Val.[2]	Cal.[1]	Val.[2]	Cal.[1]	Val.[2]
Compton Creek	0.871	0.000	0.058	0.000	1.258	0.000	0.797	0.000	0.163	0.000
Rio Hondo	No flow									
Arroyo Seco	0.000	0.005	0.000	0.315	0.000	0.239	0.000	0.015	0.000	0.015
Verdugo Wash	0.008	0.328	0.002	0.225	0.104	1.193	0.011	0.099	0.003	0.027
Western Burbank	0.200	0.000	0.677	0.000	1.967	0.000	0.010	0.000	0.155	0.000
Tujunga Wash	0.233	0.236	0.120	0.228	1.667	1.160	0.010	0.050	0.009	0.044
Bell Creek	0.056	0.091	0.293	0.120	0.373	0.328	0.003	0.003	0.051	0.048

[1] Based on data collected on September 11-12, 2000
[2] Based on data collected on July 29-30, 2001

Stormwater Water Quality

SCCWRP measured flow and water quality at 48 dry-weather stormwater inputs on the LA River and tributaries during the September 2000 data collection and at 69 dry-weather stormwater inputs on the LA River mainstem during the July 2001 data collection (Figures 14 and 15). The data collected by SCCWRP were used to assign representative flow and nutrient concentrations to each of the individual stormflows, characterized as inputs to the model cells corresponding to their measurement location. Dissolved oxygen was not measured at any of the dry-weather stormwater flows. However, a value for DO needs to be input into the WASP model. A sensitivity analysis was conducted to evaluate potential DO values and their effect on the model results. The WASP model was run with DO values of 3 to 10 mg/L for the stormwater inputs. Changing the stormwater DO value produced a minimal effect on the model results. Therefore, a value of 6.0 mg/L, corresponding to the point source permit limits, was input into the model for each stormwater flow.

3.2 Hydrodynamic Model Calibration and Validation

Model calibration is a critical component of the TMDL modeling analysis. Calibration consists of comparing model results to observed data to evaluate the accuracy of the model simulations and adjusting relevant parameters to obtain simulations that appropriately represent the behavior of the system. Once the calibration provides acceptable results, a model validation is conducted. The validation includes application of the calibrated model to a data set that is independent of the calibration data set (e.g., data from a different time period) to evaluate the ability of the calibrated model to appropriately simulate the system under different conditions or time periods.

The LA River hydrodynamic model was calibrated for low flow conditions measured on the dates of the first intensive data collection (September 10 and 11, 2000) and then validated to the flow conditions measured during the second monitoring effort on July 29-30, 2001. After the model was calibrated, model validation was performed using the data collected on July 29 and 30, 2001. The following section presents the results of the calibration of the hydrodynamic model of the LA River system.

3.2.1   Hydrodynamic Calibration (September 10 and 11, 2000)

EFDC was calibrated to observed data collected on September 10 and 11, 2000. The dataset represents a snapshot picture of the flow distribution in the LA River and the hydrodynamic model was calibrated to this snapshot by running the model under constant values of the headwater, point source discharges and stormwater inflows and reaching an equilibrium condition. The model was then adjusted to match the longitudinal distribution of the measured flow and water surface elevation. The transport was calibrated by matching the modeled velocities to those measured by SCCWRP during the September 2000 time of travel studies.

The only model parameter that was adjusted during calibration of the EFDC hydrodynamic model to alter the flow is the Manning's n value for each segment. Calibration was used to determine the final Manning's n values that were used in all subsequent simulations, including validation and evaluation of the TMDL scenarios.

Figure 17 presents a longitudinal plot of flow in the LA River, the minimum and maximum flow values measured on September 11, 2000, and the locations of the Tillman and Glendale WWRPs. The measured flows range from 50 to 120 cfs at the upper most station (mile 38) to about 135 to 200 cfs at the lowest station (mile 4). The lowest station (F319-R) is below the confluence of all tributaries in the LA River and all simulated point source discharges. Figure 17 shows that the model simulated the upper range of the measured data at three of the four stations.

During the summer of 2000, SCCWRP also performed three dye studies within the main stem of the LA River - downstream of the Bell Creek confluence, downstream of the Glendale WWRP, and near the 4th Street Bridge. The data from these dye studies were to be used in calibrating the model for velocity. Data from the dye studies indicated that at two of the sites (Bell Creek and 4th Street Bridge) there was a loss of dye from the drop point to the measurement point. This indicates that while the dye was traveling downstream, some of it escaped into the flood plain and did not reenter the low flow channel, causing inaccurate calculations of the velocity in the river. Therefore, data were not used from the Bell Creek or the 4th Street dye studies and only the results of the dye study downstream of the Glendale WWRP were used for calibration of the modeled velocity. Figure 18 presents the simulated longitudinal plot of velocity with the measured and simulated velocity at the dye study location.

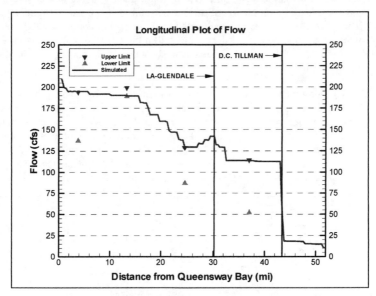

Figure 17. Simulated vs. Measured Flow During 2000 Low Flow Period

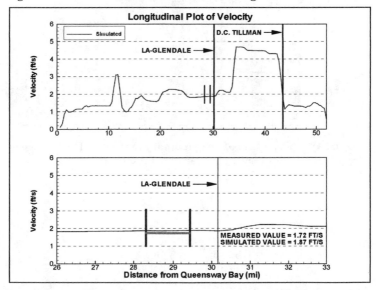

Figure 18. Simulated vs. Measured Velocity During 2000 Low Flow Period

3.3   Water Quality Model Calibration and Validation

As with the hydrodynamic model, the WASP water quality model was calibrated and validated to evaluate its simulation of the water quality response and conditions in the LA River watershed. WASP was calibrated to the snapshot conditions measured on September 10 and 11, 2000, and then validated to the snapshot conditions measured on July 29 and 30, 2001, both under quasi steady state conditions with constant loads and forcing functions.

Before performing the model calibration and validation, the 2000 and 2001 SCCWRP water quality data were evaluated to better understand the water quality conditions during the simulation periods. Figures 19 through 21 provide comparisons between the measured longitudinal distribution of nitrogen compounds within the main stem LA River between the 2000 and 2001 low flow measurements. Figure 22 presents the effluent concentrations from the Tillman and Glendale plants to provide an understanding of the relative magnitude to instream concentrations.

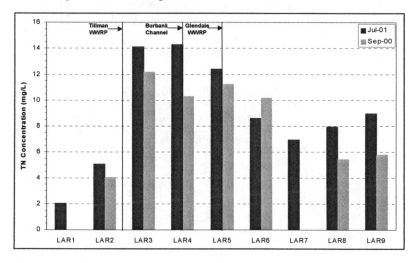

Figure 19. Comparison of Longitudinal Transects of Total Nitrogen (September 2000 vs. July 2001)

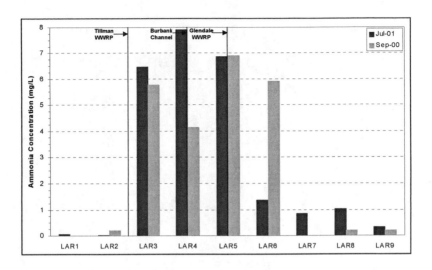

**Figure 20. Comparison of Longitudinal Transects of Ammonia
(September 2000 vs. July 2001)**

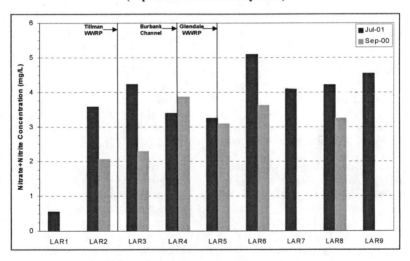

**Figure 21. Comparison of Longitudinal Transects of Nitrate/Nitrite
(September 2000 vs. July 2001)**

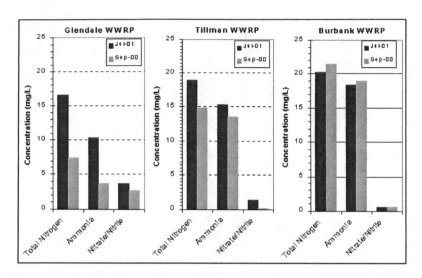

Figure 22. WWRP Effluent Concentrations for September 2000 and July 2001

Observed data indicate that the processes effecting the longitudinal distribution of the nitrogen species moving downstream from the points of discharge for the major discharges are similar for the two data collection efforts. While the 2001 data collection effort shows higher overall concentrations of ammonia and other nitrogen constituents for the two major discharges, the processes of nitrification and uptake of total nitrogen in the system appear to be similar. Both data sets show a significant reduction in the ammonia concentrations within the system in the area of LAR6 and LAR7. Additionally, both data sets show a reduction in the total nitrogen within the system moving downstream with a slight increase at the lower stations. In both data sets the lowest concentrations of total nitrogen below the discharge points is around LAR7 and LAR8, with an increase moving down to LAR9.

The data analysis identified important water quality processes and aspects to evaluate during the model calibration and validation. Although the model simulations included dissolved oxygen and carbonaceous oxygen demand, these processes were not the primary focus of the modeling effort. The shallow nature of the system and the high velocities provided sufficient reaeration to the system such that oxygen depletion was not significant. The more important processes were the nitrogen cycling relative to ammonia and nitrate/nitrite concentrations, the phosphorus cycling, and algal growth. The focus of the model calibration was to accurately capture the processes of nitrification of ammonia, uptake of nutrients by attached algal components, and dilution associated with the distribution of discharges along the system and their relative concentrations of nitrogen and phosphorus.

Water quality calibrations and validations were performed for ammonia, nitrate + nitrite, organic nitrogen and total phosphorous for both the LA River and its listed tributaries. Total nitrogen was not a calibration/validation parameter and was calculated by summing the ammonia, nitrate + nitrite and organic nitrogen concentrations. Attached algae were simulated on the LA River and were compared to measured data collected by SCCWRP (Kamer, 2000). However, the model was not calibrated for attached algae due to limited data and therefore model parameters were not adjusted from the values obtained from Warwick et al. (1997) which are consistent with SCCWRP data. The following sections present and discuss the results of the calibration of the WASP water quality model for the LA River system.

3.3.1 Water Quality Calibration (September 10 and 11, 2001)

Calibration of the WASP water quality model was conducted for September 10 and 11, 2001. Modeled results were compared to observed data for the listed tributaries as well as the mainstem of the LA River. The following sections present the calibration results for the tributary and mainstem calibrations.

3.3.1.1 Tributary Water Quality Calibration

The water quality calibrations for the tributaries are presented in Figures 23 through 26. The model results for each of the listed tributaries were calibrated to observed data at their confluence with the LA River. The measured data consisted of a single sample collected on September 11, 2000, not allowing for evaluation of temporal variability in water quality.

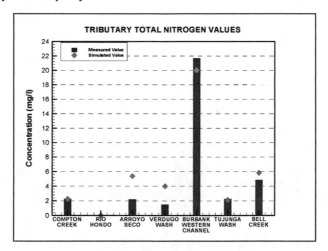

Figure 23. Simulated vs. Measured Total Nitrogen on the Listed Tributaries for the 2000 Low Flow Period

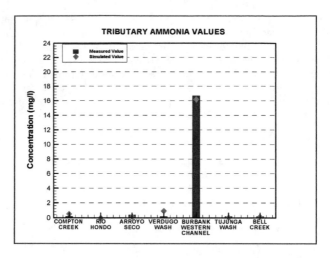

Figure 24. Simulated vs. Measured Ammonia on the Listed Tributaries for the 2000 Low Flow Period

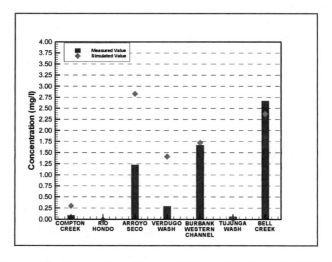

Figure 25. Simulated vs. Measured Nitrate+Nitrite on the Listed Tributaries for the 2000 Low Flow Period

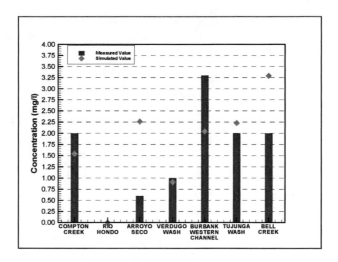

**Figure 26. Simulated vs. Measured Organic Nitrogen on the Listed
Tributaries for the 2000 Low Flow Period**

The calibration effort primarily focused on calibration of ammonia values, because ammonia is a dominant portion of the nitrogen loading in the system. As shown in Figure 26, the WASP model simulates ammonia concentrations well during the calibration period. Tributary data for attached algae were only collected on Arroyo Seco. The mean biomass measured was 1903.0 (+/- 311.6) g/m² and the simulated biomass was 38.0 g/m². The limited data from the single sampling event were used only for relative comparison to model results because of its limited spatial representation of algal biomass in the LA River system and the processes that affect it. Therefore, the model was not adjusted in an attempt to better reproduce the single observed value for biomass and it was assumed that using default values would be more appropriate.

Burbank Western Channel dominates the tributary nutrient load to the LA River due to the Burbank WWRP. Even though the other tributaries have high concentrations of total nitrogen and total phosphorous, their flow and loads are minor compared to those of Burbank Western Channel.

3.3.1.2 Los Angeles River Mainstem Water Quality Calibration

As in the tributaries, the calibration effort in the LA River mainstem primarily focused on calibration of ammonia values, because ammonia is a dominant portion of the nutrient loading in the system. For all nutrient parameters, calibration was also considered successful if the model simulated relative trends reflected in the observed data.

The water quality calibrations for the LA River are presented in Figures 27 through 32. The calibration points for the LA River consisted of seven composite samples collected along the river on September 11, 2000. The simulated water quality concentrations follow the same trend as measured water quality concentrations for all constituents except nitrate+nitrite. This is most likely due to the lack of knowledge about the nitrogen cycle below stations LAR-6.

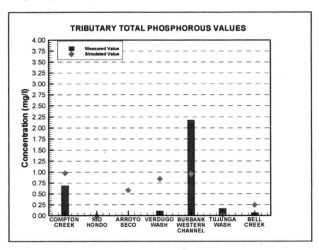

Figure 27. Simulated vs. Measured Total Phosphorous on the Listed Tributaries for the 2000 Low Flow Period

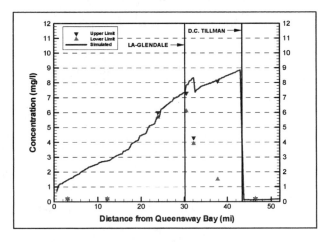

Figure 28. Simulated vs. Measured Ammonia on the LA River for the 2000 Low Flow Period

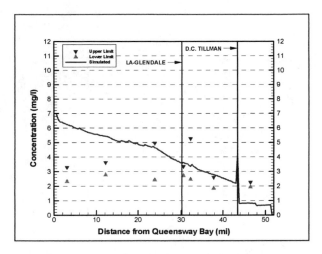

Figure 29. Simulated vs. Measured Nitrate+Nitrite on the LA River for the 2000 Low Flow Period

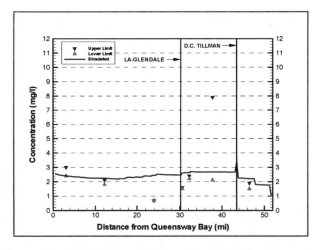

Figure 30. Simulated vs. Measured Organic Nitrogen on the LA River for the 2000 Low Flow Period

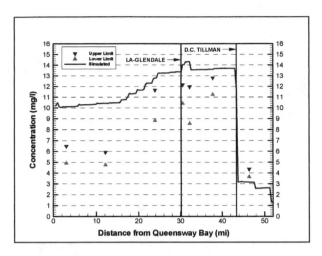

**Figure 31. Simulated vs. Measured Total Nitrogen on the
LA River for the 2000 Low Flow Period**

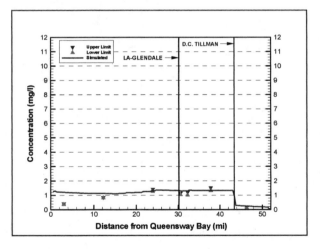

**Figure 32. Simulated vs. Measured Total Phosphorous on the
LA River for the 2000 Low Flow Period**

The seven composite samples reflect a quasi-steady state condition that provided a longitudinal pattern to target for model calibration. The key element in the calibration process was to try and capture as well as possible the overall processes occurring, this included nitrification, dilution due to additional point and non-point source discharges to the main stem, uptake by attached algae, and additional loss of total nitrogen occurring in the river below mile 20. In general the model captures the trends in the system.

For the nitrogen the nitrification of the ammonia can be seen from the point of the primary discharge (Tillman) down to around mile 20. The model appears to be capturing the overall slope in the rise of nitrate-nitrite and the fall of the ammonia. These slopes are a combination of dilution and nitrification. Examination of the total nitrogen indicates (due to the concentrations remaining constant) that the bulk of the process is nitrification down to just above mile 20. From mile 20 down to the mouth there is a significant drop in the total nitrogen concentration that, based on the water balance in the system, cannot be explained due to dilution from additional source waters entering the system. Visual inspection of the main stem of the river in this area indicates the possibility of bacterial mats that may be providing a significant uptake of nitrogen from the system. Data further support the evidence of a large uptake of nitrogen occurring within this region of the LA River main stem. Although the model does not directly capture this uptake in the lower portions, for the purposes of the nutrient TMDL the key point of compliance is the area immediately downstream of the primary discharge (Tillman) where the model does appear to be accurately capturing the overall processes of dilution and nitrification.

3.3.3   Summary of Water Quality Calibration

As shown in Figures 22 through 32, the model appears to be simulating the water quality constituents in a consistent manner. The calibration and validation results generally follow the same pattern as each other and as the observed water quality. There appears to be a large uptake of total nitrogen downstream of the Glendale WWRP (mile 30) that the model is not simulating well. This uptake of total nitrogen is most likely occurring from algae as well as the volatilization of ammonia. Because the data are not available to characterize these processes, it was assumed more appropriate to leave the model at default values rather than try to reproduce the uptake of total nitrogen reflected in the limited dataset. The model is simulating organic nitrogen and total phosphorous well in both the calibration and validation. Overall, it appears that the model is a valuable tool to predict water quality for evaluation of TMDL scenarios in the LA River.

4.   Development of TMDL

The model was used to evaluate four potential management options for reducing nitrogen loadings to the system. The first option (Scenario 1) involves nitrification and denitrification (N/DN) at the three major POTWs. Scenario 2 is based on the N/DN of Scenario 1, but evaluates the effect of 10 mgd of water reclamation at the Donald C. Tillman POTW to further reduce nitrogen loadings. Scenario 3 also involves N/DN at the major POTWs, but evaluates the effect treating 30 mgd of effluent through a constructed wetland at the Donald C. Tillman POTW. Scenario 4 is the same as scenario 3 (N/DN at the three POTWs with 30 mgd of constructed wetlands treatment) and also assumes 10 mgd of water reclamation at the Donald C. Tillman POTW.

The flow estimates are based on a reduction of plant capacity by 13 percent for N/DN facilities. The effluent quality for the N/DN process was based on estimates from pilot testing at the Los Angeles-Glendale POTW provided by the City of Los Angeles.' The effluent quality represents water quality that can be met on a monthly average. These concentrations were applied in the model to all three POTWs.

The predicted instream concentrations are presented for each of the segments of the river modeled (Table 14). The scenario evaluation assumed an effluent concentration of 2 mg/L for ammonia and 2.2 mg/L for nitrate. It is noted that the scenario evaluation utilized an ammonia load in the POTW effluent that may exceed the ammonia target for the Donald C. Tillman POTW. All four scenarios result in substantial reduction in ammonia, nitrate-nitrite and total nitrogen for the main stem and Burbank Western Channel. Under Scenario 1, total nitrogen loadings would be reduced by approximately 50 percent (from 4,375 kg/d to 2419 kg/d) over the existing condition and there would be an almost five-fold reduction of ammonia loads (from 3,328 kg/d to 722 kg/d). The 10-mgd of water reclamation would remove an additional 253 kg/d of total nitrogen from the system and the wetland option would remove an additional 602 kg/d of total nitrogen from the system.

The predicted water quality concentrations were evaluated to determine the effectiveness of each management scenario to meet the water quality objectives for ammonia and nitrate + nitrite in the Los Angeles River and tributaries along the entire length of the Los Angeles River. The model also provides output to evaluate changes in total nitrogen, phosphate, and algal biomass.

Table 15 presents a summary of the modeling results in terms of the extent of the ammonia plume concentration downstream of the Tillman WRP as a function of the ammonia as nitrogen concentration. The model indicates that the maximum instream ammonia concentration is 1.8 mg/L based on a discharge of 2.0 mg/L.

In the model, algal biomass in the Los Angeles River was not sensitive to nitrogen reduction scenarios. There was only a slight reduction in algal biomass in Burbank Western Channel. This is consistent with special studies performed by SCCWRP (Kamer, In Prep) that suggest that nitrogen may not be limiting algae in the Los Angeles River. A sensitivity analysis was run to estimate the concentration at which phosphorous became limiting in the model. Phosphorous was not limiting at concentrations as low as 0.3 mg/L. This analysis suggests that algal biomass in the Los Angeles River may be controlled by other processes, such as flow, substrate, turbidity, canopy cover, phosphorous and temperature, in addition to nitrogen concentrations.

Further research is needed to determine whether nitrogen compounds are controlling algal biomass in the river and if so what levels of reductions would be necessary to limit algal biomass. Due to this uncertainty, the implementation plan includes monitoring to observe changes in algae mass. If algal growth is not sufficiently reduced to meet targets, further analysis will be conducted to revise this TMDL for nitrogen compounds and include other pollutants that affect algal growth.

Table 14. Comparison of Flows, Nitrogen Concentrations, and Nitrogen Loadings for Four Management Scenarios to Existing Condition

Existing Condition Donald C. Tillman	Flow (mgd)	Concentrations (mg/L)				Loading (kg/d)			
		NH₃	NO₃-NO₂	Org-N	Total N	NH₃	NO₃-NO₂	Org-N	Total N
Direct Discharge	34.4	13.4	0.1	1.8	15.3	1745	13	234	1992
Japanese Gardens	4.8	12.5	0.9	3.1	16.5	227	16	56	300
Recreation Lake	17.4	4.4	7.6	4.3	16.2	286	497	283	1067
Wildlife Lake	5.9	12.5	0.9	3.1	16.5	279	20	69	368
Glendale POTW	9.3	3.7	2.7	1.0	7.4	129	95	35	259
Burbank POTW	9.2	19.0	0.5	2.0	21.5	662	17	70	749
	81.0					3328	659	748	4735
Scenario 1	Flow (mgd)	Concentrations (mg/L)				Loadings (kg/d)			
		NH₃	NO₃-NO₂	Org-N	Total N	NH₃	NO₃-NO₂	Org-N	Total N
Donald C. Tillman	70.0	2.0	2.7	2.0	6.7	530	715	530	1775
Burbank	8.0	2.0	2.7	2.0	6.7	61	82	61	203
Glendale	17.4	2.0	2.7	2.0	6.7	132	178	132	441
	95.4					722	975	722	2419
Scenario 2	Flow (mgd)	Concentrations (mg/L)				Loadings (kg/d)			
		NH₃	NO₃-NO₂	Org-N	Total N	NH₃	NO₃-NO₂	Org-N	Total N
Donald C. Tillman	60.0	2.0	2.7	2.0	6.7	454	613	454	1522
Burbank	8.0	2.0	2.7	2.0	6.7	61	82	61	203
Glendale	17.4	2.0	2.7	2.0	6.7	132	178	132	441
	85.4					646	873	646	2166
Scenario 3	Flow (mgd)	Concentrations (mg/L)				Loadings (kg/d)			
		NH₃	NO₃-NO₂	Org-N	Total N	NH₃	NO₃-NO₂	Org-N	Total N)
Donald C. Tillman	40.0	2.0	2.7	2.0	6.7	303	409	303	1014
Tillman Wetland	30.0	1.6	2.0	0.1	1.4	182	227	11	159
Burbank	8.0	2.0	2.7	2.0	6.7	61	82	61	203
Glendale	17.4	2.0	2.7	2.0	6.7	132	178	132	441
	95.4					677	895	506	1817

Table 14. Comparison of Flows, Nitrogen Concentrations, and Nitrogen Loadings for Four Management Scenarios to Existing Condition (con't)

Scenario 4	Flow (mgd)	Concentrations (mg/L)				Loadings (kg/d)			
		NH₃	NO₃-NO₂	Org-N	Total N	NH₃	NO₃-NO₂	Org-N	Total N
Donald C. Tillman	30.0	2.0	2.7	2.0	6.7	227	307	227	761
Tillman Wetland	30.0	1.6	2.0	0.1	1.4	182	227	11	159
Burbank	8.0	2.0	2.7	2.0	6.7	61	82	61	203
Glendale	17.4	2.0	2.7	2.0	6.7	132	178	132	441
	85.4					601	793	431	1564

Table 15. Magnitude (mg/L) and extent (miles) of ammonia signal downstream of Donald C. Tillman WRP under Four Nitrogen Reduction Scenarios

NH3-N concentration (mg/L)	Scenario 1	Scenario 2	Scenario 3	Scenario 4
1.8	0	0	0	0
1.7	1.88	0.75	0	0
1.6	5.26	4.13	0	0
1.5	9.37	7.52	3.75	1.88
1.4	10.81	10.11	7.89	5.26
1.3	14.37	13.27	10.86	9.75
1.2	16.57	16.20	14.73	12.62
1.1	18.41	17.51	16.94	16.20
1.0	19.14	19.14	18.77	18.04

5.   Allocations

In this section, wasteload allocations for nitrogen compounds from point sources, and allocations for nitrogen compounds from nonpoint sources to the Los Angeles River are developed. The wasteload allocations discussed below are based on Scenario 2, which was selected by stakeholders as the preferred scenario.

5.1   Wasteload Allocations

U.S. EPA regulations require that a TMDL include wasteload allocations (WLAs), which identify the portion of the loading capacity allocated to existing and future point sources (40 CFR 130.2(h)). It is not necessary that every individual point source have a portion of the allocation of pollutant loading capacity. It is necessary, however, to allocate the loading capacity among individual point sources as necessary to meet the water quality objective.

This TMDL defines ammonia WLAs in accordance with Resolution No. 2002-11 and the Policy for Implementation of Toxics Objectives for Inland Surface Waters, Enclosed Bays, and Estuaries. The ammonia Waste Load Allocation for this TMDL is equivalent to the Effluent Concentration Allowance (ECA) as defined in the Policy for Implementation of Toxics Objectives. The ECA is based on the ammonia Water Quality Objectives (WQO) and provides the basis, along with an analysis of the variability in POTW denitrification performance, for determining effluent limits for ammonia in NPDES permits. Because the dischargers have not yet implemented nitrification at the major POTWs, it is difficult to quantify the variability in nitrification performance that is necessary to determine the ammonia effluent limits. Consequently, the POTW effluent limits for ammonia necessary to implement the WLAs for this TMDL will be specified in the NPDES permit.

5.1.1 Wasteload Allocations for Major Point Sources

WLAs have been developed for the Donald C. Tillman, Los Angeles-Glendale and Burbank POTWs because they represent approximately 85 percent of the total nitrogen loadings to the system. Wasteload allocations for Donald C. Tillman, Los Angeles-Glendale and Burbank POTWs are based on concentrations needed to meet in-stream water quality objectives for ammonia, nitrate-N + nitrite-N, nitrate, and nitrite. The WLAs are set at levels necessary to attain and maintain the applicable narrative and numerical water quality objectives. A 20 percent explicit margin of safety has been included for nitrate, nitrite, and nitrate + nitrite to account for any lack of knowledge concerning the relationships between effluent limitations and water quality.

WLAs for ammonia are based on Resolution No. 2002-11 which establishes the relationship between water quality objectives and the beneficial uses of inland waterbodies. Since most of Los Angeles River listed segments are not designated in the Basin Plan as "COLD," "MIGR," and "SPWN," it is assumed that salmonids are absent and early life stages are not present in Los Angeles River. WLAs for ammonia (NH3) include one-hour and thirty day averages and are based on the pH and temperature data downstream from the POTWs for the past five years. The 90th percentile of pH data is used to establish the one-hour average WLA, and the medians of pH and temperature data are used to establish the thirty-day average WLA. WLAs for Donald C. Tillman, Los Angeles-Glendale, and Burbank POTWs are provided in Table 16. The ammonia WLA for the Donald C. Tillman WRP has been modified to account for increased assimilative capacity from discharge into the Los Angeles River that passes through the Wildlife and Recreational Lakes where ammonia is converted to oxidized nitrogen. The magnitude of the increased assimilative capacity is based on the product of a ratio of the total effluent to the effluent directly discharged through the Lakes (80 MGD/63 MGD) and an estimate of the magnitude of ammonia conversion from 2001 monitoring data. The estimate of ammonia conversion is based on the average ammonia concentration in the effluent to the average concentration in the Wildlife Lake Receiving Water Station W-3 (16.2 mg/L and 14.7 mg/L, respectively), i.e. 9 percent conversion. Therefore, WLA for ammonia at the Tillman WRP is adjusted by a factor of 1.05. If the water effect ratio study results in a revised ammonia objective, this TMDL will be revised to reflect the new ammonia target and correspondent WLA.

Table 16. Ammonia (NH₃) Wasteload Allocation for Major POTWs in Los Angeles River Watershed

POTWS	One-hour Average WLA (mg/L)	Thirty-day Average WLA (mg/L)
Donald C. Tillman WRP	4.2	1.4
Los Angeles-Glendale WRP	7.8	2.2
Burbank WRP	9.1	2.1

Table 17. Nitrate-Nitrogen, Nitrite-Nitrogen, and Nitrate-Nitrogen + Nitrite-Nitrogen Wasteload Allocations for Major POTWs

POTWs	Thirty-day Average WLA* (mg/L)		
	NitrateNO₃-N	NitriteNO₂-N	NitrateNO₃-N +NitriteNO₂-N
Donald C. Tillman WRP	7.2	0.9	7.2
Los Angeles-Glendale WRP	7.2	0.9	7.2
Burbank WRP	7.2	0.9	7.2

*Receiving water monitoring is required on a weekly basis to ensure compliance with the water quality objective

Table 17 shows the WLAs for nitrate-nitrogen (NO3-N), nitrite-nitrogen (NO2-N), and nitrate-nitrogen plus nitrite-nitrogen (NO3-N + NO2-N) for major POTWs in the Los Angeles River watershed.

These limits will be sufficient to meet the water quality objectives. This assertion is based on two key findings from the Source Analysis and Linkage Analysis. The first finding is that there are no other point sources with sufficient loads to increase nitrogen compound concentrations above the water quality objective. This finding is reasonable warranted based on the Source Analysis, however it is conceivable that this could change in the future. For this reason it may be prudent to develop wasteload allocations for the minor NPDES dischargers. This will require development of improved monitoring programs to establish the baseline from these sources. The second finding is that there are no sinks in the system that would allow for the accumulation of nitrogen. This also appears to be warranted since most of the river is channelized and sediments that may accumulate in these channels are likely to be flushed out during major storms. The one possible exception would be in the vicinity of the Glendale Narrows where willow trees and other vegetation have taken root. This area is a relatively small portion of the river and the overall effect on the nitrogen budget for the river is probably negligible.

5.1.2 Wasteload Allocations for Minor Point Sources

Ammonia WLAs for minor point sources will be set at levels necessary to maintain the applicable water quality objective. WLAs for minor point sources will be established in accordance to the reach into which a minor point source discharges based on instream pH and temperature of the last five years data set. Ammonia WLAs for minor point source discharges are listed in Table 18.

6.   DiToro, D.M., Fitzpatrick, J.J., and Thomann, R.V. 1981, rev. 1983. Water Quality Analysis Simulation Program (WASP) and Model Verification Program (MVP) - Documentation. Hydroscience, Inc., Westwood, NY, for U.S. Environmental Protection Agency, Duluth, MN, Contract No. 68-01-3872.

7.   Dodds, W.K. and E.G. Welch, 2000. Establishing nutrient criteria in streams. J. N. Am. Benthol. Soc. 19(1): 186-196.

8.   Dodds, W.K., V. H. Smith, and B. Zander, 1997. Developing nutrient targets to control benthic chlorphyll levels in streams: A case study of the Clark Fork River. Wat. Res. 31(7): 1738-1750.

9.   Duke, L. D., M. Buffleben, and L. A. Bauersachs. 1998. Pollutants in storm water runoff from metal plating facilities, Los Angeles, California. Waste Management 18:25-38.

10.   Hamrick, J.M. 1996. A User's Manual for the Environmental Fluid Dynamics Computer Code (EFDC). The College of William and Mary, Virginia Institute of Marine Science, Special Report 331, 234 pp.

11.   Kamer, In Prep. Biomass and nutrient uptake rates of filamentous green algae in the Los Angeles River. Tech. Note. Southern California Coastal Water Research Project, Westminster CA.

12.   LARWQCB, 2003. Total Maximum Daily Loads for Nitrogen Compounds and Related Effects – Los Angeles River and Tributaries. Los Angeles Regional Water Quality Control Board.

13.   LARWQCB, 1998a. Proposed 1998 List of Impaired Surface Waters (The 303(d) List). Los Angeles Regional Water Quality Control Board.

14.   LARWQCB, 1998b. Los Angeles River Watershed Water Quality Characterization.

15.   Los Angeles Regional Water Quality Control Board.

16.   LARWQCB, 1996. Water Quality Assessment and Documentation. Los Angeles Regional Water Quality Control Board.

17.   LARWQCB, 1994. Water Quality Control Plan Los Angeles Region (Basin Plan, June 13, 1994).

18.   NDEP. 1982. Water Quality Management (208) Plan for the Carson River Basin, Nevada. Nevada Division of Environmental Protection, Department of Conservation and Natural Resources.

19.   Strauss, 2002. Letter from Alexis Strauss [U.S. EPA] to Celeste Cantú [State Board], Feb. 15, 2002.).

20.   SWRCB, 2000a. Policy for Implementation of Toxics Objectives for Inland Surface.

21.   Waters, Enclosed Bays, and Estuaries. State Water Resources Control Board, Sacramento, California.

22.   SWRCB, 1988. Resolution number 88-63   Sources of Drinking Water Policy, California State Water Resources Control Board.

23.   SWRCB, 1968. Resolution number 68-16 Statement of Policy with Respect to Maintaining High Quality Water, California State Water Resources Control Board.

24.   Thomann, R.V., and Mueller, J.A. 1987. Principles of Surface Water Quality Modeling and Control. Harper & Row, New York. 644 pp.

25. Warwick, J.J, Cockrum, D., and Horvath, M.   1997. Estimating Non-Point Source Loads and Associated Water Quality Impacts. Journal of Water Resources Planning and Management.
26. Whitford, L.A., and Schumacher, G.J. 1964. "Effect of current on respiration and mineral uptake in Spirogyra and Oedogonium." Ecology, 45, 168-170.
27. UC Davis, 2002. Los Angeles River Toxicity Testing Project. May 2002.
28. U.S. EPA, 2000a. Guidance for developing TMDLs in California. EPA Region 9. January 7, 2000.
29. U.S. EPA, 2000b. Ambient water quality criteria recommendations. Information supporting the development of state and tribal nutrient criteria. Rivers and streams in nutrient ecoregion III. US Environmental Protection Agency. EPA-822-B-00-016.
30. U.S. EPA, 1999. Protocol for developing nutrient TMDLs. First Edition. US Environmental Protection Agency. EPA 841-B-99-007.
31. U.S. EPA, 1999, 1999 Update of Ambient Water Quality Criteria for Ammonia. US Environmental Protection Agency, EPA-822-R-99-014.

Nutrient Management and Seagrass Restoration in Tampa Bay, Florida: A Voluntary Program Meeting TMDL Requirements

H. S. Greening[1] and A. J. Janicki[2]

[1] Tampa Bay Estuary Program, 100 8th Avenue SE, MSI-1/NEP, St. Petersburg, FL 33701 (727) 893-2765 www.tbep.org
[2] Janicki Environmental, Inc., 1727 Dr. Martin Luther King Blvd., St. Petersburg, FL 33704 (727) 895-7722

Abstract

The Tampa Bay estuary is located on the eastern shore of the Gulf of Mexico in Florida, USA and is comprised of mixed land use. Between 1950 and 1990, an estimated 40-50% of the seagrass acreage in Tampa Bay was lost due to excess nitrogen loading and related increases in algae concentration, causing light limitation to seagrass survival and growth.

The Tampa Bay Estuary Program (TBEP), a partnership that includes three regulatory agencies and six local governments, was formed to build on the resource-based approach initiated by earlier bay management efforts. A key focus of the TBEP has been to establish nitrogen loading targets for Tampa Bay to encourage seagrass recovery. The TBEP has developed water quality models to quantify linkages between nitrogen loadings and bay water quality, and models that link water quality to seagrass goals.

Nutrient reduction targets necessary to meet water quality targets were adopted and are being addressed through a voluntary public/private consortium consisting of local, regional and state governments, industry, electric utilities and agricultural interests. In 1998, the State of Florida, with the encouragement of the TBEP partners, submitted the water quality and nutrient reduction targets adopted by the TBEP to USEPA as a Total Maximum Daily Load (TMDL) for nutrients (nitrogen) in the four major bay segments of Tampa Bay.

Data and observations from Tampa Bay indicate that initial efforts to reduce nitrogen loading and the continuing efforts of the TBEP and NMC partners are resulting in adequate water quality for the expansion of seagrasses.

TABLE OF CONTENTS

LIST OF TABLES

LIST OF FIGURES

1.   Introduction

1.1.   Background

The Tampa Bay estuary is located on the eastern shore of the Gulf of Mexico in Florida, USA. At more than 1000 km^2, it is Florida's largest open water estuary. More than 2 million people live in the 5700 km^2 watershed, with a 20% increase in population projected by 2010. Land use in the watershed is mixed, with about 40% of the watershed undeveloped, 35% agricultural, 16% residential, and the remaining commercial and mining (TBNEP 1996).

Major habitats in the Tampa Bay estuary include mangroves, salt marshes and submerged aquatic vegetation. Each of these habitats has experienced significant areal reductions since the 1950s, due to physical disturbance (dredge and fill operations) and water quality degradation, particularly impacting the seagrasses due to loss of light availability (Johansson and Greening 2000). The importance of seagrass as a critical habitat and nursery area for fish and invertebrates, and as a food resource for manatees, sea turtles and other estuarine organisms has been recognized by the Tampa Bay resource management community for several decades. In 1990, Tampa Bay was accepted into the U.S. Environmental Protection Agency's (USEPA) National Estuary Program. The Tampa Bay Estuary Program (TBEP), a partnership that includes three regulatory agencies and six local governments, has built on the resource-based approach initiated by earlier bay management efforts. Further, it has developed water quality models to quantify linkages between nitrogen loadings and bay water quality, and models that link water quality to seagrass goals. Nutrient reduction targets necessary to meet water quality targets were adopted and are being addressed through a voluntary public/private consortium consisting of local, regional and state governments, industry, electric utilities and agricultural interests.

1.2.   Impaired Water Quality Status

In 1998, the State of Florida, with the encouragement of the TBEP partners, submitted the water quality and nutrient reduction targets adopted by the TBEP to USEPA as a Total Maximum Daily Load (TMDL) for nutrients (nitrogen) in the four major bay segments of Tampa Bay. The USEPA approved this submittal, including allocations to local governments and to private sector members of the Tampa Bay Nitrogen Management Consortium collectively. As noted in the TMDL application, "The TMDL is based on an adopted five year nitrogen management strategy to 'hold the line' at existing annual nitrogen loadings to each segment of the bay in order to protect and restore seagrass meadows. The nitrogen load targets were developed for the major bay segments and not individual sources. This allows flexibility in the way the loads are controlled." (USEPA approval documents for the Tampa Bay nitrogen TMDL, 1998).

Subsequently, the Florida State legislative session produced a TMDL bill in 1999, called the Florida Watershed Restoration Act, which established the TMDL process for the state (summarized by Guillory and Sear 1999). The bill called for development and implementation of an Impaired Waters Rule, which included determination of impairment based on a set of criteria developed by the state, before calculating and allocating the amount of a pollutant which a water body may receive without violating water quality standards. The Act also stated that any TMDL calculations or allocations established prior to this act must undergo all of the rule adoption procedures identified in the bill, which meant that the Tampa Bay nitrogen TMDL adopted by USEPA was no longer recognized by the State of Florida. USEPA, however, still recognized those TMDLs adopted prior to the Florida Watershed Restoration Act including the Tampa Bay nitrogen TMDL, and the USEPA-approved TMDLs continue to be in effect.

The State of Florida's Impaired Waters Rule (Chapter 62-303, Florida Administrative Code) also included a provision that, if an existing program is deemed sufficient to achieve water quality compliance no further TMDL compliance will be required. The program must submit documentation providing "reasonable assurance" that the program's pollution control methods and techniques will result in meeting water quality criteria for the impaired water of concern. The Tampa Bay Estuary Program and its Nitrogen Management Consortium public and private partners developed a "Reasonable Assurance" document, which was accepted by the Florida Department of Environmental Protection (FDEP) in August 2002 (TBNMC 2002).

1.3.   Resource Goals and Water Quality Targets

Between 1950 and 1990, an estimated 40-50% of the seagrass acreage in Tampa Bay was lost due to excess nitrogen loading and related increases in algae concentration, causing light limitation to seagrass survival and growth. In 1980, all municipal wastewater treatment plants were required to provide Advanced Wastewater Treatment (AWT) for discharges directly to the bay and its tributaries. In addition to the significant reductions in nitrogen loadings from municipal wastewater treatment plants, stormwater regulations enacted in the 1980s also resulted in reduced nitrogen loads to the bay (Johansson and Greening 2000). Estimates for average annual total nitrogen loadings to Tampa Bay for 1976 are more than 2.5 times as high as current estimates (Zarbock et al. 1994; Pribble et al. 2001).

A key focus of the TBEP has been to establish nitrogen loading targets for Tampa Bay to encourage seagrass recovery. In 1996, local government and agency partners in the TBEP approved a long-term goal to restore 95% of the seagrass coverage observed in 1950. In 1998, the Tampa Bay Nitrogen Management Consortium (NMC) was formed. The NMC includes local governments and agencies participating in the TBEP, and phosphate companies, electric utilities and agricultural interests in the Tampa Bay watershed. These entities have pledged to work cooperatively in a voluntary, non-regulatory framework to assist with the maintenance of nitrogen loads to support seagrass restoration in Tampa Bay (Greening 2001; Greening and DeGrove 2001).

Data and observations from Tampa Bay indicate that initial efforts to reduce nitrogen loading and the continuing efforts of the TBEP and NMC partners are resulting in adequate water quality for the expansion of seagrasses. Time series plots show that, with the exception of the 1998 El Nino year, chlorophyll a targets have been met in all four major bay segments since 1994. Seagrass acreage increased an average of 350-500 acres per year between 1988 and 1996. Heavy rains associated with El Nino resulted in seagrass loss of approximately 2000 acres between 1996 and 1999 (Tomasko 2002); however, seagrass acres recorded in January 2002 show seagrass recovery in many areas of the bay where seagrass was lost between 1996 and1999 (Tomasko et al. in press).

2.   Waterbody Characteristics

Tampa Bay is located on the Gulf coast of Florida (Figure 1). Four major bay segments comprise of the mainstem Tampa Bay system: Hillsborough Bay, Old Tampa Bay, Middle Tampa Bay and Lower Tampa Bay (Figure 2). The total surface area of the four segments is 882 km^2.

Figure 1. Tampa Bay Estuary, Located on Florida's Gulf Coast

2.1. Water Use Classification

Class II and Class III, as defined by F.A.C. 62-302.400. Class III is defined as Recreation, Propagation and Maintenance of a Healthy, Well-Balanced Population of Fish and Wildlife, and applies to all portions of the waterbody.

TAMPA BAY

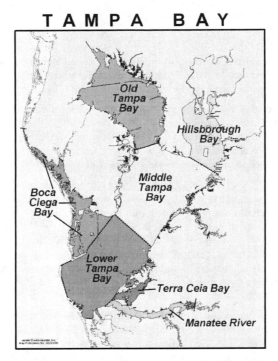

Figure 2. Four Major Bay Segments for Tampa Bay

All of Hillsborough Bay, Old Tampa Bay, Middle Tampa Bay and Lower Tampa Bay are designated for the propagation and maintenance of a healthy, well-balanced population of fish and wildlife (also referred to as "Aquatic Life Use Support" or ALUS). Several bay segments are identified by FDEP as not meeting ALUS due to nutrient impairment. Such impairment is based on monitoring chlorophyll *a* relative to generic, statewide criteria developed under Florida's Impaired Waters Rule (IWR), Chapter 63-303, F.A.C. However, all bay segments currently meet the site specific chlorophyll *a* targets established by the TBEP, which are based on many years of directed study and research within the major segments of Tampa Bay (Janicki et al. 2003).

3.   Pollutant of Concern

The pollutant of concern has been identified as Total Nitrogen, which has been determined to be the limiting nutrient in Tampa Bay. Elevated nitrogen loading has been demonstrated to lead to excess algal growth (as indicated by chlorophyll *a* concentrations), which in turn leads to reduced light penetration and loss of seagrass in the bay.

3.1 Sources of Pollutant of Concern

The sources of nitrogen loads to Tampa Bay are varied and include point sources, non-point sources, atmospheric deposition, groundwater/springs, and fertilizer losses from two port facilities (Pribble et al. 2001). Nitrogen loading estimates combine both measured nitrogen loads and estimated loads. Brief descriptions of the methods used to estimate each source type are described here and fully in Pribble et al. 2001.

The hydrologic load to the bay via precipitation was estimated using an inverse distance-squared method applied to data from 22 National Weather Service rainfall monitoring sites in the Tampa Bay watershed. Monthly rainfall estimates were used to develop direct wet deposition loads to the bay's surface, and to estimate non-point source pollutant loads from ungaged parts of the watershed.

Approximately 57% of the watershed is gaged for both flow and water quality, allowing for direct estimates of loads. For ungaged areas, loads from stormwater runoff are estimated using predictions based on rainfall, land use, soils and seasonal land-use-specific water quality concentrations. For domestic and industrial point source load estimates, values for all individual facilities with direct surface discharges and all land application discharges with an annual average daily flow of 0.1 MDG or greater are calculated from measurements of discharge rates and constituent levels required for maintaining permit compliance. These loads are then summed for all point sources (Pribble et al. 2001).

Wet atmospheric deposition of nitrogen directly to the open waters of Tampa Bay was calculated by multiplying the volume of precipitation onto the bay by nitrogen concentration in rainfall. Dry deposition was estimated using a seasonal dry:wet deposition ratio derived from five years of concurrent wet and dry deposition measurements (Poor et al. 2001).

Groundwater flows were estimated for each of the bay segments. Only groundwater inflow that entered the bay directly from the shoreline or bay bottom was considered. Groundwater and septic tank leachate inflow to streams was accounted for through measured or modeled surface water flow and was attributed as non-point source loading, and was not included in groundwater loading estimates. Wet and dry season groundwater flow estimates were calculated using a flow net analysis and Darcy's equation, following the methods of Brooks et al. (1993). Total nitrogen (TN) concentration data for the surficial, intermediate and Floridan aquifers were obtained from the Southwest Florida Water Management Ambient Ground Water Monitoring Program.

Worst-case nitrogen loads were estimated for the mid-1970s. Approximately 8,200 metric tons entered Tampa Bay annually during this period. Point sources dominated the nitrogen loads accounting for 55% of the total load. The contributions of atmospheric deposition directly to the bay's surface, non-point sources, groundwater, and fertilizer losses were 22%, 16%, 3%, and 5%, respectively.

Since the mid-1970s a number of actions were taken to address the problem of excessive nitrogen loading to Tampa Bay. First, in 1980, all municipal wastewater treatment plants were required to provide Advanced Wastewater Treatment (AWT) for discharges directly to the bay and its tributaries. AWT required TN concentrations in the wastewater discharged to the bay not exceed 3 mg/L, reducing TN loads from this source by 90%. In addition to the significant reductions in nitrogen loadings from municipal wastewater treatment plants, stormwater regulations enacted in the 1980s also contributed to reduced nitrogen loads to the bay (Johansson and Greening 2000). Lastly, the phosphate industry initiated a number of best management practices to reduce the nitrogen loads from the port facilities from which its fertilizer products are shipped. Annual TN loads have been estimated for the period 1985-1998. These management actions resulted in a significant reduction (60%) in nitrogen loading from that estimated from the mid-1970s.

Sources of nitrogen (1995-1998 average for all four bay segments combined) are shown in Figure 3 (Pribble et al. 2001). Percent contributions from major sources are:

Stormwater	62%
Direct Atmospheric Deposition	21%
Domestic Wastewater	8%
Groundwater and Springs	4%
Industrial Wastewater	4%
Fertilizer Terminal Losses	1%

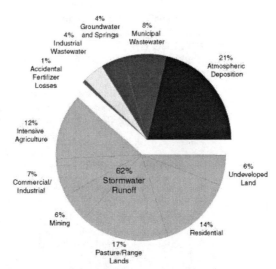

Total Nitrogen Loadings to Tampa Bay (1995-1998 average)

All Sources = 5,130 tons/year

Figure 3. Total Nitrogen Loading to Tampa Bay (1995-1998 Average). From Pribble et al. 2001).

4.   Analytical and Modeling Approaches for Tampa Bay

4.1   Technical Basis for the Goal-setting Process

Recent recommendations from the National Academy of Science National Research Council (NRC) include those that regional watershed programs might consider in developing nutrient management strategies (NRC 2000). The NRC recommendations are based on the process designed by the Tampa Bay Estuary Program partners to develop and implement a seagrass protection and restoration management program for Tampa Bay. Critical elements of the Tampa Bay process are to:

Step 1. Set specific, quantitative seagrass coverage goals for each bay segment.

Step 2. Determine seagrass water quality requirements and appropriate nitrogen loading targets.

Step 3. Define and implement nitrogen management strategies needed to achieve load management targets.

STEP 1.   SET QUANTITATIVE RESOURCE MANAGEMENT GOALS

The establishment of clearly defined and measurable goals is crucial for a successful resource management effort. The TBNEP Management Conference adopted the initial goal to increase the current Tampa Bay seagrass cover to 95% of that present in 1950 (TBNEP 1996).

Based on digitized aerial photographic images, it was estimated that approximately 16,500 ha of seagrass existed in Tampa Bay in 1950 (Lewis et al. 1991). At that time, seagrasses grew to depths of 1.5 m to 2 m in most areas of the bay. By 1992, approximately 10,400 ha of seagrass remained in Tampa Bay (Janicki et al. 1994), a loss of more than 35% since the 1950 benchmark period. Some (about 160 ha) of the observed loss occurred as the result of direct habitat destruction associated with the construction of navigation channels and other dredging and filling projects within existing seagrass meadows, and is assumed to be non-restorable through water quality management actions.

In 1996, the TBNEP adopted a bay-wide minimum seagrass goal of 15,400 ha. This goal represented 95% of the estimated 1950 seagrass cover (minus the non-restorable areas), and includes the protection of the existing 10,400 ha plus the restoration of an additional 5,000 ha (TBNEP 1996).

STEP 2.   DETERMINE SEAGRASS WATER QUALITY REQUIREMENTS AND APPROPRIATE NITROGEN LOADING RATES

Once the seagrass restoration and protection goal was established by the participants, the next steps established the environmental requirements necessary to meet the agreed-upon goal and subsequent management actions necessary to meet those requirements. Elements of this process included the following, and are more fully described in Johansson and Greening (2000) and Janicki and Wade (1996; 2001a, b).

A.   *Determine environmental requirements needed to meet the seagrass restoration goal*

Recent research indicates that the deep edges of turtlegrass (*Thalassia testudinum*) meadows, the primary seagrass species for which nitrogen loading targets are being set, correspond to the depth at which 20.5% of subsurface irradiance (the light that penetrates the water surface) reaches the bay bottom on an annual average basis (Dixon and Leverone 1995). The long-term seagrass coverage goal can thus be re-stated as a water clarity and light penetration target. Therefore, in order to restore seagrass to near 1950 levels in a given bay segment, water clarity in that segment should be restored to the point that allows 20.5% of subsurface irradiance to reach the same depths that were reached in 1950.

B.   *Determine water clarity necessary to allow adequate light to penetrate to the 1950 seagrass deep edges*

Water clarity and light penetration in Tampa Bay are affected by a number of factors, such as phytoplankton biomass, non-phytoplankton turbidity, and water color. Janicki and Wade (1996) used regression analyses, based on long-term data provided by the Environmental Protection Commission of Hillsborough County to develop an empirical model describing water clarity variations in the four largest bay segments (Old Tampa Bay, Hillsborough Bay, Middle Tampa Bay, and Lower Tampa Bay).

Water color may be an important cause of light attenuation in some bay segments; however, including color in the regression model did not produce a significant improvement in the predictive ability of the model. Results of the modeling effort indicate that, on a bay-wide basis, variation in chlorophyll *a* concentration is the major factor affecting variation in average annual water clarity (Janicki and Wade 1996).

C.   *Determine chlorophyll a concentration targets necessary to maintain water clarity needed to meet the seagrass light requirement*

Based on a technical workshop sponsored by the TBNEP in 1992, a multi-pronged (empirical and mechanistic) water quality modeling approach was used to provide a quantitative description of the relationship between nitrogen loadings and in-bay chlorophyll *a* concentrations. The developed mechanistic approach (Wang et al. 1999) and the empirical approach (Janicki and Wade 1996) each has different but re-enforcing strengths; however, the two models are made comparable by simulating the same ten-year time period (1984-94) and are driven by the same estimated hydrologic and nutrient loadings.

The empirical regression model was used to estimate chlorophyll *a* concentrations necessary to maintain water clarity needed for seagrass growth for each major bay segment (Janicki and Wade 1996). Final action taken by the TBNEP Policy Committee on June 14, 1996 adopted goals for seagrass acreage protection and restoration and segment-specific annual average chlorophyll *a* targets (8.5 µg/l for Old Tampa Bay, 13.2 µg/l for Hillsborough Bay, 7.4 µg/l for Middle Tampa Bay, and 4.6 µg/l for Lower Tampa Bay). The chlorophyll targets are easily measured and tracked through time, and are used as intermediate measures for assessing success in maintaining water quality requirements necessary to meet the long-term seagrass goal.

D.   Determine nutrient loadings necessary to achieve and maintain the chlorophyll a targets

Due principally to significant decreases in nitrogen loading as a results of increased wastewater treatment requirements starting in 1980 (Johansson and Greening 2000), conditions as of 1992-94 appear to allow an annual average of more than 20.5% of subsurface irradiance to reach target depths (i.e., the depths to which seagrasses grew in 1950) in three of the four largest bay segments (Hillsborough Bay, Old Tampa Bay and Lower Tampa Bay). Water quality in the Middle Tampa Bay segment allowed slightly less than 20.5% to target depth. Thus, a management strategy based on "holding the line" at 1992-1994 nitrogen loading rates should be adequate to achieve the seagrass restoration goals in three of the four segments. This "hold the line" approach, combined with careful monitoring of water quality and seagrass extent, was adopted by the TBNEP partnership in 1996 as its initial nitrogen load management strategy.

As an additional complicating factor, successful adherence to the "hold the line" nitrogen loading strategy may be hindered by the projected population growth in the watershed. A 20% increase in population, and a 7% increase in annual nitrogen load, is anticipated by the year 2010 (Zarbock et al. 1994; Janicki et al. 2001). Therefore, if the projected loading increase (a total of 17 U.S. tons per year) is not prevented or precluded by watershed management actions, the "hold the line" load management strategy will not be achieved.

The TBEP uses annual average bay segment chlorophyll a levels for tracking water quality targets (Janicki and Wade 1996). Maintaining chlorophyll a concentrations at target levels is expected to result in the maintenance of water clarity levels adequate to support eventual seagrass expansion to depths observed in1950, thereby ensuring that nutrient levels do not result in an imbalance in the flora or fauna of Tampa Bay (Greening 2001).

4.2.   Procedures to Determine Whether Additional Corrective Actions are Needed

In 2000, a "decision matrix" process was developed by the TBEP Technical Advisory Committee and approved by the TBEP Management and Policy Boards to help determine if seagrass goals and water quality targets are remaining "within bounds," or if management action is required to get back on track. Recommended types of management actions if the process indicates deviation from targets are also identified. This process is applied on an annual basis to determine if water clarity and chlorophyll a concentrations are remaining at or near target levels.

The process to track the status of chlorophyll-a concentrations and light attenuation involves a decision framework leading to possible outcomes dependent upon the magnitude and duration of the target exceedence (Figure 4). The decision frameworks for chlorophyll a concentration and light attenuation are identical. The process then uses a decision matrix approach to estimate the severity of the situation based on a combination of the chlorophyll *a* concentration and light attenuation outcomes (Janicki et al. 2000). The Technical Advisory Committee (TAC) and Management Board of the TBEP have adopted this process for use in status tracking.

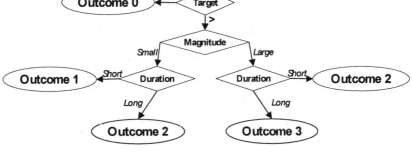

Figure 4. Monitoring and Assessment Decision Framework for Chlorophyll *a* (from Janicki et al. 2000)

As seen in Figure 4, definitions for magnitude and duration of events are needed for the process, as follows;

- Magnitude   – large vs. small
- Duration   – long vs. short.

The logic employed in deriving the approaches to defining "magnitude" and "duration" focuses on the variation in the data that have been collected by the Environmental Protection Commission of Hillsborough County (EPCHC) during the 1984-1998 period. The TBEP Water Quality Subcommittee recognized that water quality data exist from 1974 to the present, but felt the poor water quality conditions found in the bay prior to 1984 are not likely to be observed in the future, based on changes in wastewater treatment methods prior to 1984. Chlorophyll *a* concentrations are directly measured by the EPCHC, while light attenuation (K_d) is calculated from measured Secchi disc depth. The definitions for duration of difference between the ambient water quality variable value (chlorophyll *a* concentration, light attenuation) and the target value are:

- "short": < 4 years persistence, and
- "long" : ≥4 years persistence.

The definition of "short" duration is based on the minimum temporal data reporting span. Data reporting is done annually, and targets are based on segment-specific annual averages of water quality constituents. Data review and assessment with respect to targets is performed annually.

The definition of "long" duration is based upon our understanding of the response time of seagrass populations to variations in water quality conditions. This understanding is based on conversations with members of the Joint Modeling Subcommittee, which suggested that significant target exceedences found to occur during four or more consecutive years should be cause for more concern than shorter duration exceedences.

The method for defining "magnitude" is discussed below. Chlorophyll a concentration is used as the variable of interest in the development of the method, with the resultant "magnitude" definition for light attenuation following.

The approach to defining "magnitude" focuses on the EPCHC data variation, using the estimation of inter-annual variation in chlorophyll a. An estimate of the inter-annual variation was calculated by first estimating, by bay segment, the mean annual chlorophyll a concentrations. The mean annual bay segment chlorophyll a concentrations were then averaged over the entire period of record, providing a mean chlorophyll a concentration for the bay segment for the 1984-1998 period. The mean standard error was estimated for this mean 1984-1998 chlorophyll a concentration. Thus,

$$\bar{\bar{Y}} = \sum_{j=1974}^{1998} \frac{\bar{Y}_j}{n}$$

where:

$\bar{\bar{Y}}$ = mean chlorophyll a concentration for 1984-1998,

$\bar{Y}_j$ = chlorophyll a concentration in year j, and

n = number of years (15).

The inter-annual standard error for the time period of 1984 through 1998 is therefore,

$$SE_{\bar{\bar{Y}}} = \sqrt{\frac{s^2}{n}}$$

where:

$SE_{\bar{\bar{Y}}}$ = standard error of mean chlorophyll a concentration (1984-1998),

s^2 = variance in annual mean chlorophyll a concentrations (1984-1998), and

n = number of years (15).

Using these formulae, the bay segment-specific inter-annual standard errors for chlorophyll a were estimated as follows, using data from the 1984-1998 period:

Old Tampa Bay	0.4 µg/L
Hillsborough Bay	0.9 µg/L
Middle Tampa Bay	0.5 µg/L
Lower Tampa Bay	0.2 µg/L

The definitions for magnitudes of difference between the ambient chlorophyll a concentration and the chlorophyll a target concentration are based on the mean inter-annual standard error. In the listing below, the difference is denoted as O-T (observed value minus target value), and the standard error is denoted as SE. The definitions are as follow:

- O-T <1 SE   No difference in magnitude;
- 1 SE ≥ O-T ≤ 2 SE   "Small" difference in magnitude; and
- O-T >2 SE   "Large" difference in magnitude.

The "small" and "large" chlorophyll a magnitudes for each bay segment are shown below.

CHL – Magnitudes Based on Mean Inter-Annual Standard Errors		
Bay Segment : Target (μg/L)	Small Magnitude (μg/L)	Large Magnitude (μg/L)
Old Tampa Bay : 8.5	8.9 – 9.3	>9.3
Hillsborough Bay : 13.2	14.1 – 15.0	>15.0
Middle Tampa Bay : 7.4	7.9 – 8.5	>8.5
Lower Tampa Bay : 4.6	4.8 – 5.1	>5.1

The logic for defining "magnitude" described above was also applied to light attenuation as estimated from Secchi disc depth measured in the 1984-1998 period. The target light attenuation is that light attenuation that allows 20.5% of incident subsurface light to reach the target depths in each bay segment. The target light attenuation is estimated based on bay segment-specific regression parameters relating light attenuation to Secchi disc depth. The target depths, at which 95% of recoverable seagrass acreage is predicted to be achieved, are as follows (Janicki and Wade 1996):

Old Tampa Bay	1.9 m
Hillsborough Bay	1.0 m
Middle Tampa Bay	1.9 m
Lower Tampa Bay	2.5 m

The bay segment-specific inter-annual standard errors for light attenuation were estimated as follows, using data from the 1984-1998 period:

Old Tampa Bay	0.03 m^{-1}
Hillsborough Bay	0.05 m^{-1}
Middle Tampa Bay	0.04 m^{-1}
Lower Tampa Bay	0.03 m^{-1}.

The "small" and "large" light attenuation magnitudes for each bay segment are shown below, based on the inter-annual variation of light attenuation.

K_d – Magnitudes Based on Mean Inter-Annual Standard Errors		
Bay Segment: Target (m^{-1})	Small Magnitude (m^{-1})	Large Magnitude (m^{-1})
Old Tampa Bay : 0.83	0.86 – 0.88	>0.88
Hillsborough Bay : 1.58	1.63 – 1.67	>1.67
Middle Tampa Bay : 0.83	0.87 – 0.91	>0.91
Lower Tampa Bay : 0.63	0.66 – 0.68	>0.68

The magnitude and duration definitions were applied to the 1974-2001 chlorophyll-*a* data. A time series plot of chlorophyll *a* concentrations is presented for the Hillsborough Bay segment (Figure 5). The plot shows target chlorophyll *a* concentrations, small magnitude differences, and large magnitude differences.

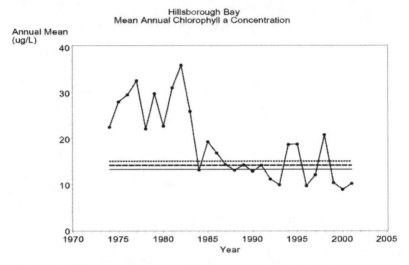

Figure 5. Hillsborough Bay Average Annual Chlorophyll *a* Concentrations, with Target (Solid Line), Small Magnitude Difference Threshold (Long Dashed Line), and Large Magnitude Difference Threshold (Short Dashed Line)

The small and large magnitude differences provided by the inter-annual variant approach were applied to the decision formulation for chlorophyll a (Figure 5), which incorporates both magnitude of difference and duration of difference. The results of this process are shown in Table 1.

Table 1. Decision Formulation Outcomes for Chlorophyll *a*.

Year	OTB Outcome	HB Outcome	MTB Outcome	LTB Outcome
1975	2	2	2	0
1976	2	2	2	1
1977	3	3	3	2
1978	3	3	3	1
1979	3	3	3	2
1980	3	3	3	2
1981	3	3	3	2
1982	3	3	3	3
1983	3	3	3	3

Table 1. Decision Formulation Outcomes for Chlorophyll *a* (con't)			
1984	2	0	3	0
1985	2	2	3	1
1986	2	2	2	0
1987	2	1	2	0
1988	0	0	0	0
1989	2	1	1	1
1990	2	0	1	0
1991	0	1	0	0
1992	0	0	0	0
1993	0	0	0	0
1994	2	2	2	2
1995	2	2	2	0
1996	0	0	0	0
1997	0	0	1	0
1998	2	2	2	2
1999	0	0	0	0
2000	0	0	0	0
2001	0	0	0	0

The magnitude and duration definitions for light attenuation were also applied to the 1974-2001 light attenuation estimated from Secchi disc depth data. A time series plot of light attenuation in Hillsborough Bay is presented in Figure 6. The plot shows target light attenuation, small magnitude differences, and large magnitude differences.

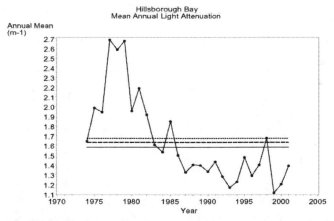

Figure 6. Hillsborough Bay Average annual Light Attenuation, with Target (Solid Line), Small Magnitude Difference Threshold (Long Dashed Line), and Large Magnitude Difference Threshold (Short Dashed Line)

The small and large magnitude differences provided by the inter-annual variant approach were applied to the decision formulation for light attenuation, which parallels that for chlorophyll a (Figure 4). The results of this process are shown in Table 2.

Table 2. Decision Formulation Outcomes for Light Attenuation

Year	OTB Outcome	HB Outcome	MTB Outcome	LTB Outcome
1975	2	2	2	0
1976	2	2	2	2
1977	2	2	3	2
1978	3	3	3	2
1979	3	3	3	3
1980	3	3	3	3
1981	3	3	3	3
1982	3	3	3	3
1983	3	0	3	3
1984	3	0	3	3
1985	3	2	3	2
1986	3	0	3	0
1987	3	0	3	0
1988	3	0	3	0
1989	3	0	3	2
1990	3	0	3	1
1991	0	0	3	2
1992	1	0	3	2
1993	2	0	3	2
1994	0	0	3	3
1995	2	0	3	3
1996	2	0	3	0
1997	2	0	3	2
1998	3	2	3	2
1999	3	0	3	2
2000	0	0	3	3
2001	2	0	2	3

The decision matrix of the status tracking process identifies appropriate categories of management actions in response to various outcomes of the chlorophyll a and light attenuation decision formulations, as shown in Table 3 (Janicki et al. 2000).

Table 3. Decision Matrix Identifying Appropriate Categories of Management Actions in Response to Various Outcomes of the Monitoring and Assessment of Chlorophyll *a* and Light Attenuation Data

CHLOROPHYLL	LIGHT ATTENUATION			
a	Outcome 0	Outcome 1	Outcome 2	Outcome 3
Outcome 0	*GREEN*	*YELLOW*	*YELLOW*	*YELLOW*
Outcome 1	*YELLOW*	*YELLOW*	*YELLOW*	*RED*
Outcome 2	*YELLOW*	*YELLOW*	*RED*	*RED*
Outcome 3	*YELLOW*	*RED*	*RED*	*RED*

The recommended management actions resulting from the decision matrix are classified by color into three categories, as follows:

- **GREEN**   "Stay the course"; partners continue with planned projects to implement the CCMP. Data summary and reporting via the Baywide Environmental Monitoring Report and annual assessment and progress reports.

- **YELLOW**   TAC and Management Board on caution alert; review monitoring data and loading estimates; attempt to identify causes of target exceedences; TAC report to Management Board on findings and recommended responses if needed.

- **RED**   TAC, Management and Policy Boards on alert; review and report by TAC to Management Board on recommended types of responses. Management and Policy Boards take appropriate actions to get the program back on track.

The results of the application of the definitions for magnitude and duration for chlorophyll *a* and light attenuation, shown in Tables 1 and 2, were applied to the decision matrix shown in Table 3. The decision matrix is based on the outcomes, as shown in Figure 4, of duration and magnitude of difference, for both chlorophyll *a* and light attenuation. This resulted in the identification of the management action categories as shown in Table 4.

Table 4. Decision Matrix Results

Year	OTB Outcome	HB Outcome	MTB Outcome	LTB Outcome
1975	Red	Red	Red	Green
1976	Red	Red	Red	Yellow
1977	Red	Red	Red	Red
1978	Red	Red	Red	Yellow
1979	Red	Red	Red	Red
1980	Red	Red	Red	Red
1981	Red	Red	Red	Red
1982	Red	Red	Red	Red
1983	Red	Yellow	Red	Red
1984	Red	Green	Red	Yellow
1985	Red	Red	Red	Yellow
1986	Red	Yellow	Red	Green
1987	Red	Yellow	Red	Green
1988	Yellow	Green	Yellow	Green
1989	Red	Yellow	Red	Yellow
1990	Red	Green	Red	Yellow
1991	Green	Yellow	Yellow	Yellow
1992	Yellow	Green	Yellow	Yellow
1993	Yellow	Green	Yellow	Yellow
1994	Yellow	Yellow	Red	Red
1995	Red	Yellow	Red	Yellow
1996	Yellow	Green	Yellow	Green
1997	Yellow	Green	Red	Yellow
1998	Red	Red	Red	Red
1999	Yellow	Green	Yellow	Yellow
2000	Green	Green	Yellow	Yellow
2001	Yellow	Green	Yellow	Yellow

5.   IMPLEMENTATION

5.1.   Participating Entities

Members of the Tampa Bay Estuary Program Policy Board include the following:

> City of Tampa
> City of Clearwater
> City of St. Petersburg
> Manatee County
> Hillsborough County
> Pinellas County
> Florida Department of Environmental Protection
> Southwest Florida Water Management District
> U.S. Environmental Protection Agency

The Tampa Bay Nitrogen Management Consortium included the following public and private entities in 2004:

Public Partners:
In addition to the nine TBEP Policy Board entities, public participants in the NMC include:

> Manatee County Agricultural Extension Service
> Environmental Protection Commission of Hillsborough County
> Tampa Bay Regional Planning Council
> Florida Fish and Wildlife Commission/Florida Marine Research Institute
> U.S. Army Corps of Engineers
> Tampa Port Authority
> Florida Department of Agriculture and Consumer Services

Private Partners:
> Florida Phosphate Council
> Florida Power & Light Company
> Tampa Electric Company
> Florida Strawberry Growers Association
> IMC-Phosphate Company
> Cargill Fertilizer, Inc.
> CF Industries, Inc.
> Pakhoed Dry Bulk Terminals (now Kinder-Morgan)
> Eastern Associated Terminals Company
> CSX Transportation

The Tampa Bay Estuary Program government partners executed an Interlocal Agreement in 1998, pledging to assist in meeting the goals of the TBEP Comprehensive Conservation and Management Plan (TBNEP 1998). Also in 1998, public and private members of the Tampa Bay Nitrogen Management Consortium pledged to exercise their best efforts to implement, either individually or in cooperation with other Consortium members, the projects they have offered to undertake as part of the Consortium Action Plan (TBNMC 1998). Many of these projects have already been completed.

5.2. Management Activities

Over 100 existing and proposed activities are included in the Tampa Bay Nitrogen Management Consortium Action Plan (TBNMC 1998). They include the following types of projects:

> Stormwater facilities and upgrades
> Land acquisition and protection
> Wastewater effluent reuse
> Air emissions reduction
> Habitat restoration
> Agricultural BMPs
> Education/public involvement
> Industrial treatment upgrades

NMC partners are currently updating projects in the Consortium Action Plan, which is being developed as an electronic database for 2001-2005 projects.

To ensure consistency, the Consortium Action Plan Database program includes a standardized method for electronically calculating both existing conditions (no treatment) TN, TP and TSS loading for each project, and estimated loadings after treatment is applied. Each treatment type (for example, wet retention pond) has been assigned a treatment efficiency based on best available data/information (Zarbock and Janicki 1997), and is applied within the database program to estimate the nitrogen load attenuation. Parameters included in these calculations are land use, soils, rainfall and hydrologic connectivity. The difference between the "treatment" and "no treatment" estimates is the load reduction anticipated for each activity. NMC partners may also propose site-specific load reduction estimates for specific projects, providing adequate documentation is provided.

5.3.   Future Growth

The TN load reduction target of 17 tons per year needed to maintain TN loading at 1992-1994 levels assumes growth in population and the associated changes in stormwater, atmospheric deposition and point sources (Janicki et al. 2001).

The TBEP Interlocal Agreement requires that the technical basis for estimating loads and establishing targets be reexamined every 5 years. The first five-year re-examination was complete in 2001. Results from the re-examination indicate that the models and assumptions used for the initial calculations continue to provide appropriate estimates of loading and resulting chlorophyll a concentrations (Janicki et al. 2001).

6.   Monitoring and Reporting Results

6.1.   Water quality monitoring programs

Existing water quality monitoring programs include ambient programs conducted by the Environmental Protection Commission of Hillsborough County, Manatee County, Pinellas County and the City of Tampa (summarized in Pribble et al. 2002). Water quality samples from over 100 stations baywide are collected and analyzed on a monthly basis through the collective efforts of these monitoring programs.

All these programs and their laboratories have State-approved Quality Assurance Plans on file, and comply with FDEP's QA rule, Chapter 62-160, including FDEP approved Standard Operating Procedures. The participating laboratories have or are working to receive NELAC certification.

Water quality reporting is done annually. In addition, TBEP conducts a full revision and update of nitrogen loading estimates (current and estimated future loads) and model evaluations every 5 years.

6.2. Implementing Management Activities

The Consortium Action Plan Database will allow entry of new projects and summary queries at any time. The TBEP staff will solicit information on new projects (or revisions to existing projects) every 2 ½ years, and will enter this information into the Database. In addition, a NMC partner can request to revise an existing project or submit a new one at any time. A formal reporting of management activities by TBEP will take place every 5 years, to correspond with the model assumption re-evaluation and CCMP update. TBEP staff is responsible for Action Plan Database maintenance.

6.3.   Evaluating Progress Toward Goals

Progress toward water quality targets is evaluated annually by the application of the "decision matrix" (Janicki and Pribble 2002). Progress towards seagrass acreage goals is evaluated every 2-3 years using the Southwest Florida Water Management District's seagrass aerial photography and digital mapping (Tomasko et al. in press).

6.4.   Proposed Corrective Actions

The "decision matrix"(Janicki et al. 2000) outlines a process by which potential management actions may be determined. In this process, the magnitude and duration of deviations from chlorophyll a and light targets are used to help determine the degree of the management response. Responses range from "green" (if all targets are met); to "yellow", in which the TAC and Management Board review monitoring data and loading estimates and attempt to identify causes of target exceedences; to "red" for cases where magnitude and duration are large and a response appears necessary. Responses to "yellow" and "red" conditions will vary according to the specific conditions of the exceedences. The Management and Policy Boards will take actions they deem to be appropriate.

7.   1995-2002 Progress

Progress to date for the TBNMC Action Plan, tracking of chlorophyll a concentration targets in the four bay segments, and baywide seagrass extent trends are summarized here.

7.1.   Tampa Bay Nitrogen Management Consortium Action Plan: 1995-2000

The types of nutrient reduction projects included in the Consortium's Nitrogen Management Action Plan (TBNMC 1998) range from traditional nutrient reduction projects such as stormwater treatment upgrades, industrial retrofits and implementation of agricultural best management practices to actions not primarily associated with nutrient reduction, such as land acquisition and habitat restoration projects. A total of 105 projects submitted by local governments, agencies and industries are included in the 1995-2000 Plan; 95% of these projects address nonpoint sources and account for 71% of the expected total nitrogen reduction. Half (50%) of the total load reduction will be achieved through public sector projects, and 50% by industry.

A total of 134 tons per year reduction in nitrogen loading to Tampa Bay is expected from the completed projects, which exceeds the 5-year reduction goal of 85 tons per year by 60%. Chlorophyll a concentrations were met in all four bay segments in 2000, 2001 and 2002, indicating that nitrogen loading is not exceeding target levels.

Examples of specific projects and expected nitrogen loading reductions in the 1995-1999 Consortium Action Plan include the following:

Stormwater facilities and upgrades: Stormwater improvements or new facilities include both public and private examples. Stormwater retrofits using alum injection to urban lakes reduced total nitrogen (TN) loading by an estimated 6.4 tons per year. Stormwater improvements eliminated an estimated 2 tons of TN loading per year. Industrial stormwater improvements at phosphate fertilizer factories and transport terminals are expected to have reduced annual TN loads by almost 20 tons per year by the year 2000.

Land acquisition and protection: Land acquisition and maintenance of natural or low intensity land uses precludes higher-density development and higher TN loadings. Land acquisition precluded more than 15 tons TN loading per year by the end of 1999. Approved zoning overlay districts requiring additional nutrient control in management areas precluded an estimated 10 tons per year.

Wastewater effluent reuse: Wastewater reuse programs resulted in a 6.4 ton per year reduction in annual TN loading. Conversion of septic systems to sewer reduced TN loading by an estimated 1.7 tons per year.

Atmospheric emissions reduction: Reductions of atmospheric emissions from coal-fired electric generating plants between 1995-1997 resulted in estimated reductions of NO_x emissions of 11,700 - 20,000 tons. To estimate the reduction of nitrogen deposition which reaches the bay (either by direct deposition to the bay's surface, or by deposition and transport through the watershed), a 400:1 ratio (NO_x emissions units to nitrogen units entering the bay) is assumed. Expected reductions from atmospheric deposition thus ranged from 29 to 50 tons per year by 1999. To date, emissions reductions have not been included in the estimated total TN reduction to the bay, pending agreement on estimation methods.

Habitat restoration: Although typically conducted for reasons other than nutrient reduction, habitat restoration to natural land uses reduces the amount of TN loading per acre via stormwater runoff. Habitat restoration projects have been completed or are underway in all segments of Tampa Bay's watershed. Estimated TN load reduction from completed habitat restoration projects totaled an estimated 7 tons per year.

Agricultural BMPs: Water use restrictions have promoted the use of microjet or drip irrigation on row crops (including winter vegetables and strawberries) and in citrus groves. Micro-irrigation has resulted in potential water savings of approximately 40% or more over conventional systems and an estimated 25% decrease in fertilizer applied. Nitrogen reduction estimates from these actions total 6.4 TN tons per year.

Education/public involvement: For those projects for which nitrogen load reductions have not been calculated or measured, but some reductions are expected, the Consortium Action Plan assumes a 10% reduction estimate until more definitive information is available. These programs have reduced TN loading by an estimated 2 tons per year.

Industrial upgrades: A phosphate fertilizer mining and manufacturing plant has terminated the use of ammonia in flot-plants (a mineral separation process), resulting in a reduction of 21 tons per year of nitrogen loading. Other fertilizer manufacturing companies have upgraded their material handling systems, resulting in a TN reduction of more than an estimated 10 tons per year due to control of fertilizer product loss. The termination of discharge by an orange juice manufacturing plant into a tributary of Tampa Bay has resulted in a reduction of more than 11 tons per year TN loading.

Ongoing efforts to create an electronic database of existing projects (such as those summarized above for the 1995-2000 time period) will include proposed projects and estimated load reductions through the year 2005.

7.2.   Chlorophyll *a* Targets

Results of applying the Decision Matrix to 2002 water quality data show that average annual chlorophyll *a* concentration targets are being met in all bay segments. Time series plots for chlorophyll *a* show that, with the exception of the 1998 El Nino year, chlorophyll *a* targets have been met in all four bay segments between 1994 and 2001 (Figure 7).

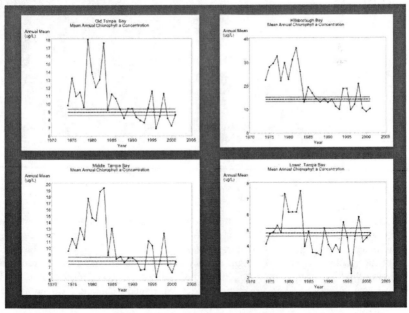

Figure 7. Time Series of Average Annual Chlorophyll *a* Concentrations in Tampa Bay's Four Major Bay Segments, All Stations Combined. Dotted Line Represents the Adopted Chlorophyll *a* Concentration Target for that Bay Segment. Solid Lines Represent ± 1 SD

7.3.   Seagrass Goals

The TBEP adopted goal for seagrass coverage is to recover an additional 11,920 acres of seagrass over 2002 levels, while preserving the bay's existing 26,080 acres. Between 1988 and 1996, seagrass acreage increased at about 350 acres per year. El

Nino rains resulted in seagrass losses of about 2,000 acres between 1996-1999. In January 2002, seagrass acreage increased by 1,237 acres baywide, a 5% increase from 1999 (Figure 8).

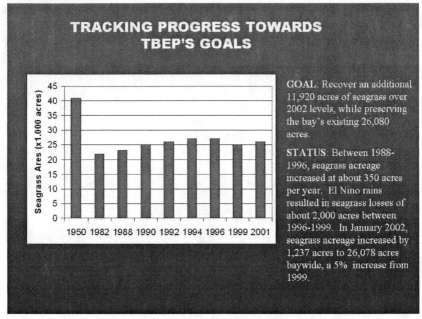

Figure 8. Seagrass Coverage (Acres) in Tampa Bay, 1950 - 2001.
Sources: Lewis et al. 1991; Tomasko et al. In Press (1988 – Present).

SUMMARY

Participants in the Tampa Bay Estuary Program have agreed to adopt nitrogen loading targets for Tampa Bay based on the water quality and related light requirements of the seagrass species *Thalassia testudinum* (turtlegrass). Based on modeling results, it appears that light levels can be maintained at necessary levels by "holding the line" at existing nitrogen loadings. However, this goal may be difficult to achieve given the 20% increase in the watershed's human population and associated 7% increase in nitrogen loading that are projected to occur over the next 10-20 years.

To address the long-term management of nitrogen sources, a Nitrogen Management Consortium of local electric utilities, industries and agricultural interests, as well as local governments and regulatory agency representatives, has developed a Consortium Action Plan to address the target load reduction needed to "hold the line" at 1992-1994 levels. To date, implemented and planned projects collated in the Consortium Action Plan meet and exceed the agreed-upon nitrogen loading reduction goal. Although not originally developed for this purpose, the Tampa Bay voluntary nutrient reduction program has been accepted by USEPA and the State of Florida as meeting TMDL requirements for nutrients in Tampa Bay.

8. REFERENCES

1. Brooks, G.R., T.L. Dix, and L.J. Doyle. 1993. Groundwater/surface water interactions in Tampa Bay and implications for nutrient fluxes. Tampa Bay National Estuary Program Technical Publication #06-93. St. Petersburg, FL.

2. Dixon, L. K. and J.R. Leverone. 1995. Light requirements of *Thalassia testudinum* in Tampa Bay Florida. Final Report. Mote Marine Laboratory Technical Report 425. Mote Marine Laboratory, Sarasota, Florida.

3. Greening, H. 2001. Nutrient Management and Seagrass Restoration in Tampa Bay, Florida, USA. InterCoast; Fall 2001.

4. Greening, H. and B.D. DeGrove. 2001. Implementing a Voluntary, Nonregulatory Approach to Nitrogen Management in Tampa Bay, FL: A Public/Private Partnership. IN: Optimizing Nitrogen Management in Food and Energy Production and Environmental Protection: Proceedings of the 2nd International Nitrogen Conference on Science and Policy. TheScientificWorld; 1(S2), 378-383.

5. Guillory, B. and T. Sear. 1999. The Florida Watershed Restoration Act and TMDL Program. Florida Water Resources Journal, September 1999, pages 29 and 33.

6. Janicki, A. and R. Pribble. 2002. Tracking Chlorophyll-a and Light Attenuation in Tampa Bay: Application to 2001 Data. Technical Report #03-02 of the Tampa Bay Estuary Program. St. Petersburg, FL.

7. Janicki, A., R. Pribble, and M. Winowitch. 2003. Tracking Chlorophyll-a and Light Attenuation in Tampa Bay: Application to 2002 Data. Technical Report #01-03 of the Tampa Bay Estuary Program, St. Petersburg, FL.

8. Janicki, A., R. Pribble, H.Zarbock, S. Janicki, and M.Winowitch. 2001. Model-Based Estimates of Total Nitrogen Loading to Tampa Bay: Current Conditions and Updated 2010 Conditions. 2001.Technical Report #08-01 of the Tampa Bay Estuary Program. St. Petersburg, FL.

9. Janicki, A. and D.Wade. 1996. Estimating Critical Nitrogen Loads for the Tampa Bay Estuary: An Empirically Based Approach to Setting Management Targets. 1996. Technical Publication #06-96 of the Tampa Bay National Estuary Program, St. Petersburg, FL.

10. Janicki, A. and D.Wade. 2001a. Tampa Bay Estuary Program Model Evaluation and Update: Chlorophyll a-Light Attenuation Relationship. Technical Report #06-01 of the Tampa Bay Estuary Program. St. Petersburg, FL.

11. Janicki, A. and D. Wade. 2001b. Tampa Bay Estuary Program Model Evaluation and Update: Nitrogen Load-Chlorophyll a Relationship. 2001.Technical Report #07-01 of the Tampa Bay Estuary Program. St. Petersburg, FL.

12. Janicki, A., D.Wade and J.R. Pribble. 2000. Developing and Establishing a Process to Track the Status of Chlorophyll-a Concentrations and Light Attenuation to Support Seagrass Restoration Goals in Tampa Bay. 2000. Technical Report #04-00 of the Tampa Bay Estuary Program. St. Petersburg, FL.

13. Janicki, A., D. Wade, and D.. Robison. 1994. Habitat Protection and Restoration Targets for Tampa Bay. Technical Publication #07-93 of the Tampa Bay National Estuary Program. St. Petersburg, FL.

14. Johansson, J.O.R. and H.S. Greening. 2000. Seagrass Restoration in Tampa Bay: A Resource-Based Approach to Estuarine Management. Pages 279-294. IN: Subtropical and Tropical Seagrass Management Ecology (S.A. Bortone, ed.), Boca Raton, FL. CRC Publication.

15. Lewis, R.R., K. D. Haddad, and J.O.R. Johansson. 1991. Recent areal expansion of seagrass meadows in Tampa Bay: Real bay improvement or drought-induced? p. 189-192. In Treat, S.F. and P.A. Clark (eds.), Proceedings, Tampa Bay Area Scientific Information Symposium 2. Text, Tampa, Florida.

16. National Research Council. 2000. Clean Coastal Waters: Understanding and Reducing the Effects of Nutrient Pollution. Committee member, Committee on the Causes and Management of Coastal Eutrophication. Ocean Studies Board and Water Science and Technology Board, National Research Council. 405 p. National Academy Press, Washington, D.C.

17. Poor, N., R. Pribble and H. Greening. 2001. Direct wet and dry deposition of ammonia, nitric acid, ammonium and nitrate to the Tampa Bay Estuary, FL, USA. Atmospheric Environment 35 (2001) 3947-3955.

18. Pribble, R., A. Janicki, H. Zarbock, S. Janicki and M. Winowitch. 2001. Estimates of Total Nitrogen, Total Phosphorus, Total Suspended Solids, and Biochemical Oxygen Demand Loadings to Tampa Bay, Florida. 2001. Technical Report #05-01 of the Tampa Bay Estuary Program, St. Petersburg, FL.

19. Pribble, R., A. Janicki, and H. Greening, (eds) 2002. Tampa Bay Baywide Environmental Monitoring Report. 2002. Technical Report #06-02 of the Tampa

Bay Estuary Program. Prepared by the Tampa Bay environmental monitoring groups. St. Petersburg, FL.

20. Tampa Bay National Estuary Program. 1996. Charting the Course: The Comprehensive Conservation and Management Plan for Tampa Bay. St. Petersburg, FL.

21. Tampa Bay National Estuary Program. 1998. Interlocal Agreement to ensure that the Tampa Bay CCMP is properly and effectively implemented, February 1998.

22. Tampa Bay Nitrogen Management Consortium. 1998. Partnership for Progress: The Tampa Bay Nitrogen Management Consortium Action Plan 1995-1999. Tampa Bay Estuary Program, St. Petersburg, Florida.

23. Tampa Bay Nitrogen Management Consortium. 2002. Tampa Bay Watershed Management Summary. Submitted to FDEP as Reasonable Assurance, August 2002. Available from the Tampa Bay Estuary Program.

24. Tomasko, D.A. 2002. Status and trends of seagrass coverage in Tampa Bay, with reference to other southwest Florida estuaries. Pages 11- 20 IN: H. Greening (ed). Seagrass Management: It's Not Just Nutrients! Proceedings of a Symposium. August 22-24, 2000, St. Petersburg, Florida.

25. Tomasko, D.A., C.A. Corbett, H.S. Greening, and G.A. Raulerson. In press. Spatial and Temporal Variation in Seagrass Coverage in Southwest Florida: Assessing the Relative Effects of Anthropogenic Nutrient Load Reductions and Rainfall in Four Contiguous Estuaries. Mar. Poll. Bull. In press.

26. U.S. Environmental Protection Agency. 1998. Approval documents for the Tampa Bay nitrogen TMDL.

27. Wang, P.F., J. Martin and G. Morrison. 1999. Water Quality and Eutrophication in Tampa Bay, Florida. Estuarine, Coastal and Shelf Science 49: 1-20.

28. Zarbock, H., A. Janicki, D. Wade, D. Heimbuch, and H. Wilson. 1994. Estimates of total nitrogen, total phosphorus, and total suspended solids loadings to Tampa Bay, Florida. Technical Publication #04-94 of the Tampa Bay National Estuary Program. St. Petersburg, Florida.

29. Zarbock, H.W. and A. J. Janicki. 1997. Guidelines for Calculating Nitrogen Load Reduction Credits. Publication #02-97 of the Tampa Bay Estuary Program. St. Petersburg, Florida.

Nutrient and Siltation TMDL Development for Wissahickon Creek, Pennsylvania

Stephen Carter[1], Tom Henry[2], Rui Zou[3], and Leslie Shoemaker[4]

[1] Tetra Tech, Inc., 10306 Eaton Place, Suite 340, Fairfax, VA 22030 (703) 385-6000
[2] United States Environmental Protection Agency – Region 3, 1640 Arch St., Philadelphia, PA 19103-2029
[3] Tetra Tech, Inc., 10306 Eaton Place, Suite 340, Fairfax, VA 22030, (703) 385-6000
[4] Tetra Tech, Inc., 10306 Eaton Place, Suite 340, Fairfax, VA 22030, (703) 385-6000

Abstract

As the result of biological investigations conducted by the Pennsylvania Department of Environmental Protection (PA DEP) that identified observed impacts on aquatic life and numerous exceedances of the applicable dissolved oxygen (DO) criteria, much of the Wissahickon Creek watershed has been listed on the State's 303(d) list of impaired waters. The watershed is heavily impacted by urbanization and is listed as impaired due to problems associated with elevated nutrient levels, low dissolved oxygen concentrations, siltation, water/flow variability, and habitat alterations. This study fulfilled the requirements for nutrient and siltation TMDL development for all waters in the Wissahickon Creek basin included in the State's 303(d) list. For those stream segments listed as impaired as a result of "water/flow variability" and "other habitat alterations," sources of impairments are related to those sources contributing to the nutrient and siltation impairments. Therefore, through implementation of best management practices to address nutrient and siltation TMDLs, these related impairments are addressed indirectly.

TABLE OF CONTENTS

LIST OF TABLES

LIST OF FIGURES

1.0   Introduction

As the result of biological investigations conducted by the Pennsylvania Department of Environmental Protection (PA DEP) that identified observed impacts on aquatic life and numerous exceedances of the applicable dissolved oxygen (DO) criteria, much of the Wissahickon Creek watershed has been listed on the State's 303(d) list of impaired waters. The watershed is heavily impacted by urbanization and is listed as impaired due to problems associated with elevated nutrient levels, low dissolved oxygen concentrations, siltation, water/flow variability, and habitat alterations. This study fulfilled the requirements for nutrient and siltation TMDL development for all waters in the Wissahickon Creek basin included in the State's 303(d) list. For those stream segments listed as impaired as a result of "water/flow variability" and "other habitat alterations," sources of impairments are related to those sources contributing to the nutrient and siltation impairments. Therefore, through implementation of best management practices to address nutrient and siltation TMDLs, these related impairments are addressed indirectly.

The Wissahickon Creek drains approximately 64 square miles and extends 24.1 miles in a southeasterly direction through lower Montgomery and northwestern Philadelphia Counties (Figure 1) of Pennsylvania. Major tributaries in the basin include Sandy Run and Pine Run, draining a heavily urbanized area east of the mid-section of the watershed. Other tributaries to Wissahickon Creek include Trewellyn Creek, Willow Run - East, Willow Run - West, Rose Valley Tributary, Paper Mill Run, Creshiem Creek, Monoshone Creek, Prophecy Creek, Lorraine Run, Wises Mill Tributary, and Valley Road Tributary. All tributaries mentioned are included with the mainstem of the Wissahickon Creek on Pennsylvania's 303(d) list of impaired waters. The headwaters and upper portions of the watershed consist primarily of residential, agricultural, and wooded land use. The mid-section of the watershed is dominated by industrial, commercial, and residential land use. The lower 6.8 miles of the watershed is enclosed by Fairmount Park, which is maintained for recreational use. Tributaries of the lower portion of the watershed provide storm drainage from single and multi-family residential areas.

PA DEP biological investigations of Wissahickon Creek over the past 20 years have repeatedly documented a problem regarding eutrophic conditions in the mainstem and tributaries (Boyer, 1975; Strekal, 1976; Boyer, 1989; Schubert, 1996; Boyer, 1997; Everett, 2002). Total phosphorus concentrations decreased substantially in 1988 as a result of a combination of the phosphate ban and wastewater treatment plant upgrades and/or phasing out of smaller treatment plants. However, levels are still significant enough to result in nuisance algal growth (Boyer, 1997). Results of a 1998 survey of the periphyton conducted by PA DEP indicate that excess nutrient levels in the Wissahickon Creek may be contributing to impairments found in the watershed by causing an alteration in the benthic community as a result of increasing algal biomass (Everett, 2002). Analysis of the periphyton data by the Academy of Natural Sciences of Philadelphia (ANSP) concluded that the Wissahickon Creek is a nutrient enriched system, with eutrophic conditions present in the stream as a whole. ANSP further concluded that this eutrophication can be attributed to sewage treatment plant (STP) effluents and possibly leached fertilizers and other runoff (West, 2000; Everett,

2002). As further evidence of eutrophic conditions, diurnal dissolved oxygen sampling performed by PA DEP in 1999 and 2002 showed repeated violations of State water quality criteria.

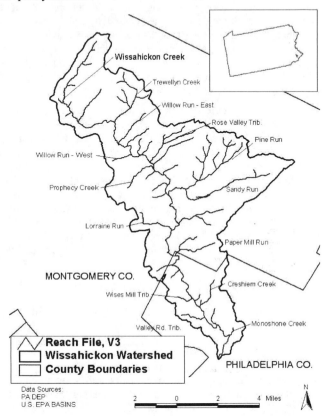

Figure 1. Wissahickon Creek Watershed

Ten stream segments in the Wissahickon Creek watershed have been included in Pennsylvania's 303(d) list of impaired waters due to nutrient impairments. These include five segments of the Wissahickon Creek mainstem as well as five stream segments of tributaries. Although nutrients are required to support a healthy biological assembly, excessive nutrient loading can be detrimental to the biological system. Excessive nutrients fosters an unhealthy and expanded growth in primary production which decreases DO levels in the stream when these organisms respire in evening hours or when they are broken down by bacterial agents upon completion of their life-cycle. Sources of nutrients have been identified as municipal point sources and urban runoff/storm sewers.

Twenty-one stream segments in the Wissahickon Creek watershed have been included on Pennsylvania's 303(d) list due to siltation impairments. These include the six segments of Wissahickon Creek as well as fifteen additional stream segments in the watershed. Excessive sediment loading and siltation are detrimental to the biological community for many reasons. Siltation reduces the habitat complexity through the filling of pools and interstitial spaces between gravel and sand. Excess sediment can clog an organism's gill surfaces, which decrease its respiratory capacity. This pollutant also impacts visual predators by negatively impacting their ability to hunt and feed in a more turbid environment. Sources of siltation impairments include urban runoff/storm sewers and habitat modification.

Another impact on the biological community and a source of impairment is the diminution of baseflow. Several portions of the headwaters and tributaries have exhibited no baseflow during PA DEP 1997 inspections conducted in conjunction with the Unassessed Waters Program, an August 2001 site visit conducted by PA DEP and EPA Region 3, and PA DEP data collection of Summer 2002. Sources of baseflow reduction may be a result of one or more of several activities, including the increase of impervious area and subsequent loss of groundwater recharge resulting from urbanization, and groundwater pumping and drawdown. Diminution of baseflow is addressed directly as an impairment included in the 303(d) list under the category of Water/Flow Variability. Management practices recommended in this study to address nutrient and siltation impairments also address impairments due to Water/Flow Variability.

Habitat alteration is affected not only by increased biomass and diminution of baseflow, but also hydraulic/hydrology changes resulting from increased urbanization. Generally, there are three major forms of habitat modification related to hydrologic/hydraulic enhancements caused by urbanization: (1) instream modifications produced by increased stormflows (siltation, bank destabilization, embeddedness, etc.), (2) out-of-stream habitat alterations (riparian vegetation removal, bank alteration, etc.), and (3) stream encroachments (dams, enclosures, bridges, etc.). All three categories of habitat modification are interrelated and are addressed directly as a source of impairment for segments included in the 303(d) list for Habitat Alterations. Siltation and Water/Flow Variability are also addressed separately in the 303(d) list, but are related to Habitat Alterations. Since they are related to the same source of impairment, the management practices identified to relieve the nutrient and siltation impairments will have a positive impact on the habitat alteration impairments as well.

2.0   Watershed Characterization

A wide range of data and information were reviewed for potential use in the development of a nutrient and siltation TMDLs for the Wissahickon Creek watershed. The categories of data examined include physiographic data describing physical conditions of the watershed, environmental monitoring data identifying potential pollutant sources and contributions to streams, hydrologic flow data, and in-stream water quality monitoring data. Table 1 shows the various data types and data sources reviewed and collected for the Wissahickon Creek watershed.

Table 1. Inventory of Data for the Wissahickon Creek Watershed

Data Category	Description	Data Source(s)
Weather Data	Rainfall	NCDC[1]
	Air Temperature	NCDC[1]
Streamflow Data	USGS Streamflow Gage	USGS[2]
	2002 Streamflow Field Measurements	PADEP
Instream Water Quality Data	Water Quality Monitoring Data	USEPA
	1998 Instream WQ Sensor Data	PADEP
	1998 WQ Field Measurements	PADEP
	1996 and 1998 WQ Field Measurements	PADEP
	1998 Biological Assessment Data	PADEP
	1999 Diurnal DO Data	PADEP
	2002 WQ Field Measurements	PADEP
Watershed Physiographic Data	Land Use	MRLC[3]
	Stream Reach Coverage	USGS[2]
	Stream Cross-Sections	ANSP
	Wissahickon watershed	USGS[2]
Environmental Monitoring Data	NPDES Data	PADEP, USEPA
	303(d) Listed Water	PADEP

2.1 Hydrology

There are 5 US Geological Survey (USGS) gages providing historical streamflow records in the Wissahickon Creek watershed (Figure 2). Table 2 summarizes the streamflow data for each gaging station. The only active streamflow gages are on the mainstem of Wissahickon Creek at the mouth (01474000) and at Fort Washington (01473900); all other gages were decommissioned twenty years ago or more. However, gage 01473900 was out of service from March 1969 to June 2000, providing little more than a year of recent data relevant for use. Since a large amount of the urbanization of the Wissahickon Creek watershed has occurred in more recent times, only the active gages are believed to provide an accurate record of the effects of urbanization on Wissahickon Creek flow. Water quality data is available at each of the USGS gage locations except USGS 01473870.

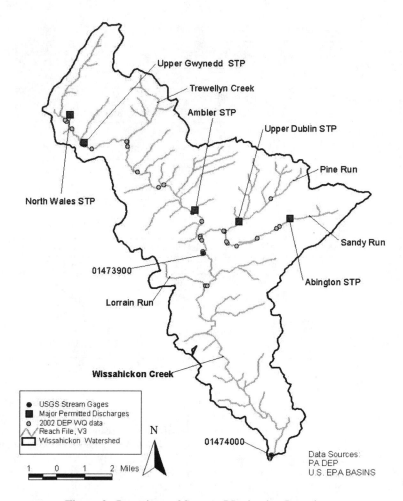

Figure 2. Locations of Stream Monitoring Locations

Table 2. USGS Stream Gages Within the Wissahickon Creek Watershed

Station	Stream Name	Start Date	End Date	Drainage Area (sq. mi.)	Min. (cfs)	Max. (cfs)	Average (cfs)
01474000	Wissahickon at Ridge Ave. Bridge	10/1/1965	present	64	8.8	5560	103.9
01473980	Wissahickon at Livezey Lane	10/1/65	11/3/1970	59.2	8.5	2060	67.9
01473950	Wissahickon at Bells Mill Road	10/1/1965	9/30/1981	53.6	7.9	2390	83.2
01473900[1]	Wissahickon at Fort Washington	9/1/1961	present	40.8	4.6	1990	47.0
01473870	Pine Run near Ambler	10/1/1973	9/30/1978	1.18	0	85	1.5

[1] USGS 01473900 was discontinued on 3/69, but reactivated 6/00; historical record is not continuous.

In general practice, flows at ungaged locations of a stream are often estimated using streamflow records at other locations in close proximity, preferably on the same stream. The streamflow record at the gaged location would be normalized by dividing the flows by the drainage area at that site, resulting in a flow per unit area of the watershed. To estimate flows at a different location, the drainage area at that location would be multiplied by the normalized flow/area record for the nearest observed flow. This method would assume a 1:1 ratio between different sites. To check this assumption, gages 1474000 and 1473900 flow records were divided by their respective drainage areas, and a regression analysis was performed on the resulting flow/area records. Figure 3 shows the result of the analysis. Results of the regression analysis suggest a strong correlation between datasets and a relationship not too different than a 1:1 ratio. Therefore, simply adjusting the flow record by the difference in drainage areas and deducting point source contributions is sufficient in estimating flows at ungaged locations of the Wissahickon Creek. However, baseflow diminution has been an ongoing problem in many of the tributaries, a characteristic that will likely be unaccounted for if flows are estimated using a gage on the mainstem.

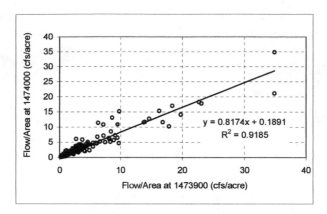

Figure 3. Wissahickon Creek Streamflow Analysis

2.2   Water Quality

Nutrient data has been collected by various agencies at multiple locations on Wissahickon Creek and its tributaries. However, the only historical record of nutrients that extends to present is at the mouth of Wissahickon Creek. From an analysis of streamflow data from USGS gage 01474000 combined with streamflow and water quality data from PADEP gage WQN0115 (Figure 2), relations between the magnitude of streamflow and levels of nutrients were established. To ensure that the analysis provides an accurate description of current conditions, data was limited to the period of record from 1990 to 2001. Figures 4 to 7 depict results from the analysis and show that levels of nitrate (NO3-N) and total phosphorus (TP) are higher during periods of low streamflows. This correlation suggests that the critical condition is during low-flow, when nutrient contributions are dominated by point sources or other direct discharges.

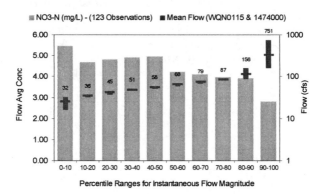

Figure 4. Analysis of Nitrate at the Mouth – Flow Variability

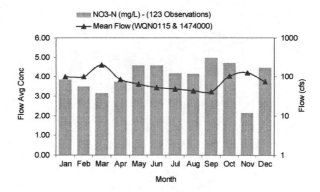

Figure 5. Analysis of Nitrate at the Mouth – Seasonal Variability

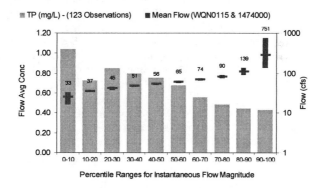

Figure 6. Analysis of Total Phosphorus at the Mouth – Flow Variability

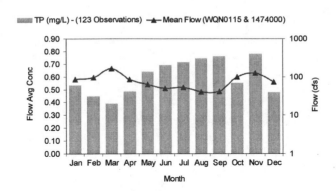

Figure 7. Analysis of Total Phosphorus at the Mouth – Seasonal Variability

Nutrient data is also observed to exhibit longitudinal variation in the streams. Data collected by PADEP in Summer 2002 revealed trends in the nutrient levels observed at different distances along Wissahickon Creek length and a major impaired tributary, Sandy Run. Figures 8 and 9 show these trends for nitrate; Figures 10 and 11 show results for total phosphorus. On this sampling date, nutrient levels appear highly influenced by the discharge of treated effluent from three wastewater treatment plants (WWTP) on Wissahickon Creek and one WWTP on Sandy Run.

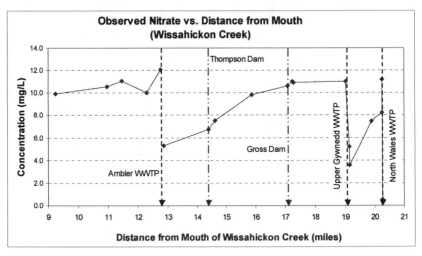

Figure 8. NO3-N Concentrations in Wissahickon Creek (Summer 2002)

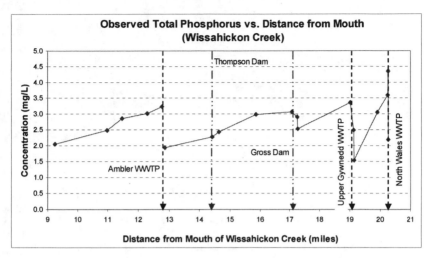

Figure 9. TP Concentrations in Wissahickon Creek (Summer 2002)

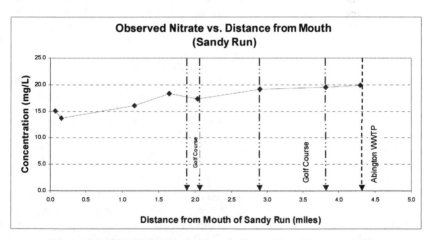

Figure 10. NO3-N Concentrations in Sandy Run (Summer 2002)

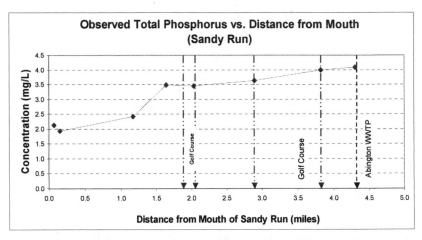

Figure 11. TP Concentrations in Sandy Run (Summer 2002)

Conclusions from these analyses strongly suggest that point sources dominate the degree of nutrient enrichment in Wissahickon Creek during the low-flow summer period. Previous biological investigations completed by PADEP have noted that as nutrient levels increase, biological activity has been observed to increase, which is the cause of observed excess periphyton growth and likely the source of the significant diurnal DO fluctuations observed in 1999. Although nonpoint sources likely contribute a noticeable amount of nutrients to the waters, the higher streamflows associated with these periods of runoff are not correlated with higher concentrations of nutrients, which suggests that flows during runoff periods are sufficient to assimilate and dilute this additional loading. Although the waters are not listed specifically as being impaired regarding low DO, data has shown repeated violations of the minimum DO standard as a result of relatively large diurnal fluctuations in DO as a result of biological activity. Such biological activity occurs during low-flow periods when high nutrient levels have been shown most common.

Historical records provide sufficient information regarding the long-term seasonal and flow-related statistics, but the sensitivity of DO levels to other biological, spatial, and hydraulic characteristics often prove more useful in analyzing its sensitivity to the environment. As mentioned, diurnal DO fluctuations are a result of biological activity that is enhanced by many factors, but namely the degree of nutrient enrichment is a major influence. Diurnal DO fluctuations are illustrated for various locations in Wissahickon Creek and Sandy Run (major tributary) in Figures 12 and 13, respectively, using data collected in Summer 2003 by PADEP. Shown in these figures are the minimum and maximum of the data collected at various locations in the stream, as well as other locations where a single DO measurement was taken. The diurnal ranges of DO data vary depending on location in the stream. DO standards were violated repeatedly at various locations throughout the basin. Problematic areas appear to be in the Wissahickon Creek headwaters, tributaries, and in the mid-section of the mainstem downstream of a major WWTP and the confluence with Sandy Run.

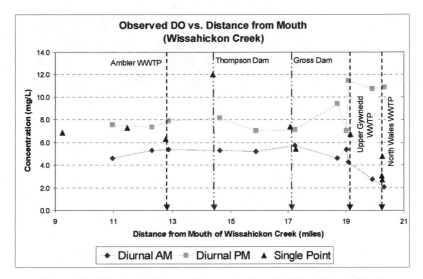

**Figure 12. Dissolved Oxygen Concentrations in Wissahickon Creek
(Summer 2002)**

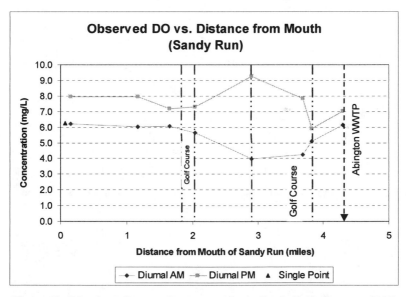

Figure 13. Dissolved Oxygen Concentrations in Sandy Run (Summer 2002)

As a result of mixing with the aerated effluent, DO levels are actually observed at some locations to increase just downstream of point sources. However, degradation of DO is observed to decrease as a function of distance downstream. DO levels at the bottom portion of Wissahickon Creek are observed to stabilize well above the minimum DO standard, with variability in data significantly less than upper portions of the headwaters and tributaries. This improvement in water quality is likely the result of the protection provided by Fairmount Park and the presence of a series of small dams that provide re-aeration.

Sources of siltation are generally associated with wet weather streamflows. To test this assumption for Wissahickon Creek, total suspended solids (TSS) levels measured at the mouth from 1990 to 2001 were compared against flows. Results of this analysis are reported in Figures 14 and 15. As can be seen from these results, TSS levels during high flows are almost an order of magnitude greater than levels observed at normal flows. Periods of such high flows and corresponding high TSS concentration suggests a relatively large solids loading and potential for siltation to the Wissahickon Creek streambed during wet periods.

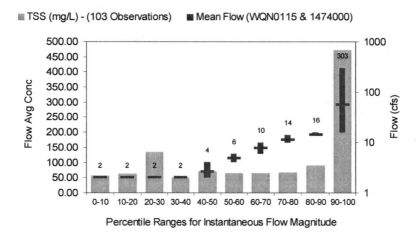

Figure 14. Analysis of Total Suspended Solids at the Mouth – Flow Variability

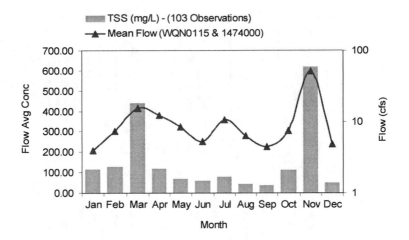

**Figure 15. Analysis of Total Suspended Solids at the Mouth –
Seasonal Variability**

3.0 Source Assessment

Analyses were performed on historical water quality and streamflow data to determine critical flow conditions and relative loads to assess the impact of point and nonpoint sources on instream water quality. These analyses helped to assess nutrient and siltation sources in the Wissahickon Creek watershed. Identification of critical flow conditions was an important step in determining the methodology used for TMDL development. Under these conditions, the relative impacts of nutrients and siltation sources differed.

3.1 Nutrient Sources

Analysis of water quality data determined that the critical period associated with impacts from high nutrient levels was during the low-flow, summer period. During low-flow periods, Wissahickon Creek nutrient concentrations are dominated by point source contributions. National Pollutant Discharge Elimination System (NPDES) permitted dischargers in the Wissahickon Creek watershed are summarized in Table 3. The discharges range from single family discharges (about 400 to 700 gallons per day) to large industrial and municipal wastewater treatment plants with effluent rates in the range of 1 to 7 million gallons per day (MGD). Major dischargers are defined in U.S. EPA NPDES Permit Writers' Manual as those facilities with design flows greater than one million gallons per day and facilities with EPA/State approved industrial pretreatment programs.

Table 3. Point Sources of Nutrient in the Wissahickon Creek Watershed

NPDES No.	Receiving Waterbody	Flow (MGD)	Facility Name	Industry Classification
PA0012190	Wissahickon Creek	0.01775	Precision Tube Co – Mueller St	Roll, Draw & Extrud Nonferrous
PA0023256	Wissahickon Creek	5.7[a]	Upper Gwynedd Township	Sewerage Systems
PA0026603	Wissahickon Creek	6.5	Ambler Boro	Sewerage Systems
PA0052515	Wissahickon Creek	0.0168	Ambler Borough Water Department	Filter Backwash From STP
PA0053538	Wissahickon Creek	na	Merck & Company, Inc	Pharmaceutical Preparations
PA0055387[d]	Wissahickon Creek	0.001	PA Historical & Museum Commission	Sewerage Systems
PA0022586	Tributary to Wissahickon Creek	0.835	North Wales Boro	Sewerage Systems
PA0054577	Tributary to Wissahickon Creek	0.0007[c]	Fishbone, David	Sewerage Systems
PA0057177[d]	Tributary to Wissahickon Creek	0.0004	Plummer, J. Randall	Sewerage Systems
PA0057576	Tributary to Wissahickon Creek	0.0007	Bruce K. Entwisle	Sewerage Systems
PA0053074	Sandy Run	0.0083	Valley Green Corporate Center	Oper of Nonresidential buildings
PA0056901	Sandy Run	0.0136	Jiffy Lube International, Inc	Auto Serv, Exc Rep & Carwashes
PA0026867	Sandy Run	3.91	Abington Township	Sewerage Systems
PA0050865	Rose Valley Tributary	0.053	Gessner Products Co Inc	Plastics Products, NEC
PA0029441	Pine Run	1.1[b]	Upper Dublin Township	Sewerage Systems
PA0013048	Pine Run	na	Honeywell, Inc.	Industrial instruments
PA0051012	Lorraine Run	0.0004	Harris, Albert & Cynthia	Oper of dwelling other than apartment
PA0057631	Lorraine Run	0.0005	Sayers, David & Marie	Sewerage Systems
PA0053210	Lorraine Run	0.0005	Murray SRSTP	Sewerage Systems

a - Approval granted 3/12/20028 for plant expansion from 4.5 to 5.7 MGD
b - Approval granted 9/18/1998 for plant expansion from 1.0 to 1.1 MGD
c - Permit expired; renewal expected
d - Permit expired; renewal questionable
na - not applicable; monitoring only

In the Wissahickon Creek watershed these facilities constitute a majority of the streamflow in the Wissahickon Creek basin during low-flow periods. Major NPDES facilities in the Wissahickon Creek basin include Ambler Borough (6.5 MGD), Upper Gwynedd Township (5.7 MGD), Abington Township (3.91 MGD), Upper Dublin Township (1.1 MGD), and North Wales Borough (0.835 MGD). Locations of all major and minor discharges are depicted in Figure 16.

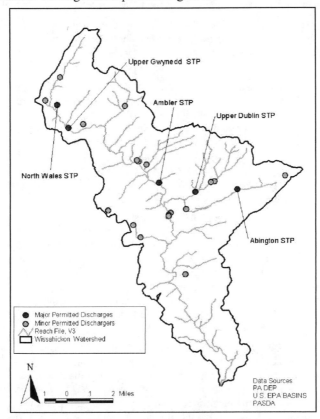

Figure 16. Locations of Point Sources in the Wissahickon Creek Watershed

During the critical low-flow period, impacts from nonpoint sources are limited since storm runoff is not a factor during such dry conditions. However, other nonpoint sources can potentially impact the streams under such conditions, including runoff from irrigated golf courses, areas with high concentration of septics and/or history of failure, unimpeded cattle access to streams, and impacts of low level dams.

During the summer 2002 instream monitoring study performed by PA DEP during low-flow critical conditions, water quality samples were taken upstream and downstream of two golf courses on Sandy Run selected to represent impacts of golf courses on streams of the Wissahickon Creek basin. If substantial impacts were observed, more robust monitoring would be performed to better characterize loads from these areas. However, during the monitoring period, no outstanding increases in nutrient concentrations were observed in the vicinity of the golf courses (Figures 10 and 11). Although increases in diurnal variability of DO in these areas (Figure 13) suggests an increase in biological activity, this occurrence is likely the result of reduced shading from tree canopy and nutrient loads from upstream sources.

PA DEP determined that during low-flow conditions, impacts from failed septic systems are negligible since most of the watershed utilizes sanitary sewer services. As a result, nutrient loads from failed septic systems were not considered in TMDL analysis.

Unimpeded cattle access is limited to one farm, but this area only impacts the lower portion of the watershed where water quality is less problematic. Moreover, without sufficient supporting data, it is difficult to make assumptions for loads from such sources. However, it was found that by reducing loads in the upstream portions of the watershed to improve conditions in the stream segments where the sources originate, the water quality improved to the point that no local reductions were required for the bottom portion of the Wissahickon Creek watershed (below Route 73). In any case, restoration projects are currently proposed by PA DEP for this portion of the watershed that will seek to reduce these impacts.

Low level dams located throughout the watershed provide opportunity for instream sources of nutrients through sediment release from pooled areas. To assess the impacts from these dams, PA DEP monitored water quality upstream and downstream of two dams on Wissahickon Creek (Figures 8 and 9). If impacts proved significant, a more robust assessment of nutrient loads from the dams would be considered. However, except for a small increase in total phosphorus at one of the dams (Gross Dam), impacts were determined minimal. Rather than attribute a source of nutrients to dams, the effects were accounted for in the water quality calibration of the model used for TMDL development.

Coorson's Quarry discharges an average of 12.5 cfs to Lorraine Run. This flow is a significant contributor to Wissahickon Creek baseflow and provides reductions to Wissahickon Creek nutrient concentrations by increasing the assimilative capacity of both Lorraine Run and the mainstem of Wissahickon Creek during the critical low-flow period. To assess the benefits of the quarry discharge, a sensitivity analysis was performed using the low-flow model. Results showed that if quarry discharges are discontinued, additional DO problems will likely result in the bottom portions of Wissahickon Creek below Lorraine Run. Also, due to the substantial reduction of streamflow that would occur in Lorraine Run, aquatic life within the stream would be affected beyond problems associated with low DO. Therefore, the discharge from Coorson's Quarry benefits the Wissahickon Creek and Lorraine Run, and continued operation of the quarry should be encouraged. This TMDL is based on the assumption that this discharge will continue its operation. If the discharge is reduced to below 0.5 cfs or terminated, the TMDL may need to be revised.

Although low-flow conditions are dominated by point source contributions, a small amount of baseflow is present with background nutrient concentrations likely controlled by groundwater. These background contributions are extremely small in comparison to point source contributions during low-flow conditions. As a result, background nutrient loads are accounted for in analyses, but impacts are negligible.

3.2   Siltation Sources

As a result of analysis of historical water quality data collected at the mouth of Wissahickon Creek, the critical condition for siltation TMDL development was during wet conditions associated with storm runoff from urban areas. Runoff from urban areas carries significant loads of sediment that deposits in the streambed. EPA's stormwater permitting regulations at 40 CFR 122.26 require municipalities to obtain NPDES permit coverage for all storm water discharges from municipal separate storm sewer systems (MS4s). Implementation of these regulations are phased such that large and medium sized municipalities were required to obtain stormwater NPDES permit coverage in 1990 (Phase I) and small municipalities by March 2003 (Phase II). As such, Philadelphia has an existing Phase I MS4 permit and surrounding smaller municipalities in the watershed were required to have NPDES Phase II MS4 permit coverage by March 2003. Figure 17 depicts the municipal boundaries within the Wissahickon Creek basin. For each municipality, the sediment loads from stormwater collection systems are considered as point source contributions, which require specific wasteload allocations in the TMDL for each MS4 permitee.

To assess the relative loads of sediment from different land uses within municipal boundaries, EPA used land use specific, unit area loadings. In order to accurately assess the loading based on this methodology, it was paramount to use the most recent and updated land use data available. A current land use dataset for the Wissahickon Creek watershed was developed by the Environmental Resources Research Institute of Penn State University by updating the National Land Cover Data (NLCD) (Vogelmann et al., 1998) using SPOT (System Probatoire pour l'Observation de la Terre) satellite imagery from 2000. The relative areas for each land use in the Wissahickon Creek basin are listed in Table 4. The most predominant land uses in the basin are low-intensity residential (38.7%), deciduous forest (26.0%), and a mix between high-density residential and urban (11.5%). Urban and residential land uses in the Wissahickon Creek basin account for over 50% of the total area, and are considered to be major contributors to sediment loads in the Wissahickon Creek watershed.

The largest contributors of sediment to Wissahickon Creek are instream sources attributed to streambank erosion. Urbanization and paving of large areas of the watershed result in dramatic increases in stormwater runoff, which lead to periodic high flows that directly cause the erosion of stream banks, contributing silt to the shallow creek bottom. These sources are extremely difficult to pinpoint, measure, and control, but they are currently the leading cause of siltation in the Wissahickon Creek basin. The cause of the flow variability that results in streambank erosion is related to urban runoff and the sources of the impairments are therefore considered point sources under the MS4 stormwater permits.

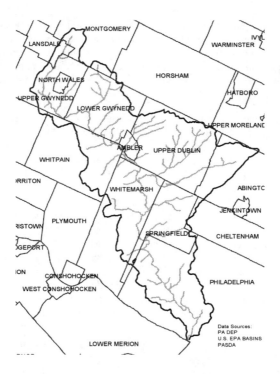

Figure 17. Municipal Boundaries in the Wissahickon Creek Watershed

Table 4. Land uses of the Wissahickon Creek watershed

Land Use	Area (sq. mi.)	Percent
Water	0.1	0.2
Low-Intensity Residential	24.7	38.7
High-Intensity Residential/Urban	7.4	11.5
Hay/Pasture	3.8	6.0
Row Crops	3.9	6.0
Coniferous Forest	1.4	2.2
Mixed Forest	5.6	8.7
Deciduous Forest	16.6	26.0
Quarry	0.2	0.2
Coal Mines	0.0	0.0
Transitional	0.2	0.4

4.0   Modeling Approach

Separate methodologies were utilized for determination of nutrient and siltation TMDLs. Each selected methodology considers specific impacts and conditions determined necessary for accurate source representation and system response.

4.1   Nutrient Modeling Approach

Results from data analyses describe the low-flow critical period associated with high observed nutrient concentrations causing low DO and harming aquatic life. To determine a TMDL for Wissahickon Creek, a low-flow, steady-state model was utilized that included chemical and biological processes associated with nutrient enriched and eutrophic systems. A steady-state model was used to simulate conditions most likely occurring during a constant, low-flow scenario typical of periods when previously observed problems are prevalent and most critical. This low-flow, steady-state model inherently focused on point sources as the major source of nutrients to the Wissahickon Creek basin. Other potential sources (i.e., runoff from golf course irrigation, impacts from low-level dams, etc.) were assessed on a case-by-case basis, but no quantitative evidence justified the inclusion of such sources in the model under such low-flow conditions.

For nutrient TMDL development, two models were utilized to simulate the hydrodynamics and water quality of the basin. EPA's Environmental Fluid Dynamics Code (EFDC) (Hamrick, 1992) was used to simulate hydrodynamics. The EFDC model is a general purpose modeling package for simulating three dimensional flow, transport, and biogeochemical processes in surface water systems including rivers, lakes, estuaries, reservoirs, wetlands, and coastal regions. The EFDC model was originally developed at the Virginia Institute of Marine Science for estuarine and coastal applications and is considered public domain software. To model water quality, a modified version of EPA's Water Quality Analysis Simulation Program (WASP5) (Ambrose et al., 1991) used results from the hydrodynamic model to simulate those processes associated with nutrients, DO, and biological activity. Modifications to the WASP5 model included sub-routines accounting for biological processes associated with periphyton growth to account for impairment effects from algal growth. This version was configured by Hydraulic and Water Resources Engineers, Inc. (HWRE) as a subcontractor to Tetra Tech, Inc. for EPA Region 1 and Maine Department of Environmental Protection, and was refined by Tetra Tech, Inc. to provide accurate adaptation to Wissahickon Creek. Both EFDC and WASP5 have been applied successfully in numerous applications to rivers, lakes, and coastal waters, and are well-known and well-documented tools for mechanistically simulating the processes of concern in Wissahickon Creek.

The model was configured for calibration to conditions observed during summer 2002 when PADEP performed instream monitoring. For this period, the flow balance for Wissahickon Creek and its tributaries was obtained through analysis of the following data:

- streamflow data during the period from July 11, 2002 to August 11, 2002 (collected by PA DEP);

- time variable discharge data provided by major dischargers for the same period; and
- streamflow data for two USGS stations located in the Wissahickon Creek basin: one at the mouth of the Wissahickon Creek main stem (USGS 01474000) and the other on Wissahickon Creek at Fort Washington (USGS 01473900).

Ultimately, the flow rate at key locations was estimated for each date during the summer 2002 low flow condition using adjustments based on respective drainage areas. The final flow distribution for the average summer low-flow condition was then determined by averaging the flow at each upstream location and from each discharger over the sampling period. The resulting flow distribution for model configuration used for calibration to summer 2002 conditions is depicted in Figure 18. During this period, flows from point sources accounted for 46.2 percent of the total Wissahickon Creek flow.

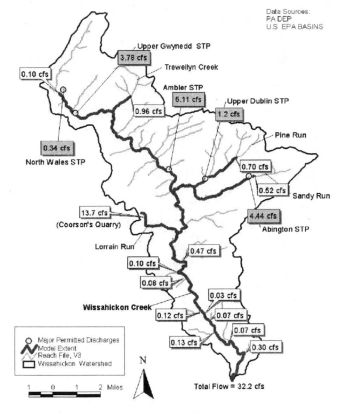

Figure 18. Summer 2002 Flow Distribution for Model Configuration

In July and August 2002, PA DEP conducted a dye study to analyze the time-of-travel in Wissahickon Creek, Pine Run, and Sandy Run during low flow conditions. The information obtained from the dye study was used to calibrate the hydrodynamic model's simulation capability. The dye study conducted by PA DEP covers a range of start-end location pairs or "schemes." A scheme represents results of time-of-travel measurements from fluorometer stations that tracked dye concentrations following injection at upstream locations. Four schemes successfully captured peaks of dye concentrations to provide sufficient information regarding time-of-travel for Pine Run, Sandy Run, and two segments of Wissahickon Creek. Locations of the schemes are depicted in Figure 19. The dye injection and flourometer station locations were mapped to the EFDC grid cell system to allow a direct comparison of model results versus the observed data.

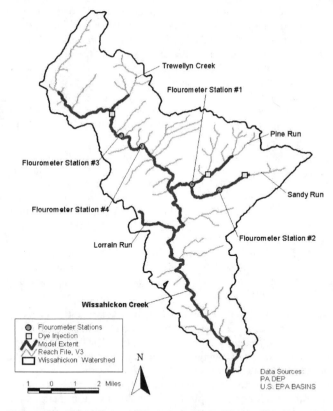

Figure 19. Wissahickon Creek Dye Study Monitoring Locations

Predicted velocities for each model segment between dye injection stations and flourometer stations were output from the hydrodynamic model. Using the velocity and the length of each segment, the time-of-travel was calculated and compared with the measured data in the dye study. In situations where the model-simulated time-of-travel deviated significantly from the observed data, the model's bottom friction coefficient (in terms of roughness height) and bathymetry were adjusted within a reasonable range.

In addition to the time-of-travel comparison, a comparison of model predicted and observed temperatures was performed to provide additional confidence in model performance. The temperature data used for calibration was collected by PA DEP during the summer 2002 survey. Since temperature is sensitive to the atmospheric boundary conditions (i.e., solar radiation and air temperature), it is advisable to compare the simulated temperature at a specific location to data collected on respective days. Note that temperatures were measured on different dates for different locations, therefore, the model calibration must consider the temporal variability of temperature dynamics. In this study, the initial temperature condition was set to be 20° C at the starting date, July 1, 2002. Since water temperature typically shows a significant diurnal fluctuation during the low flow season, the model-simulated hourly temperatures were output and used to calculate daily average, daily minimum, and daily maximum values. The simulated temperatures were compared with the observed data, and necessary adjustments to the channel roughness and bathymetry were made until a reasonable match between the model results and data was achieved.

Figure 20 shows the model-data comparison for time-of-travel. Figure 21 shows the temperature comparison results. As shown, the time-of-travel and temperatures simulated by the model correlate well with the observed data, particularly considering that the hydrodynamic model was set up for an average flow condition, while the dye study was conducted on a specific day. Moreover, since the flow condition during the survey period was relatively steady, with coefficients of variation (standard deviation divided by the mean) of 0.27 (for the flow at USGS 01473900 station) and 0.22 (for the flow at USGS 01474000), the average flow condition during the low-flow survey period was considered representative of the flow condition of the period. Successful simulation of the time-of-travel and temperature enabled the application of a water quality model for the low-flow regime.

Past investigation shows that periphyton is likely a significant contributor to the DO fluctuation in Wissahickon Creek and its tributaries due to its abundance in the creek (Tetra Tech Inc., 2002). Because the standard WASP/EUTRO model does not have a system compartment for periphyton, a modified version of the model was implemented. This modified version is capable of representing major periphyton kinetics including growth, photosynthesis, respiration, and grazing/non-grazing associated death. Similar to phytoplankton, the metabolism of periphyton is affected by environmental conditions such as temperature, nutrient limitation, and light intensity. The following equation can be used to describe periphyton dynamics (Shanaha and Alam, 2001):

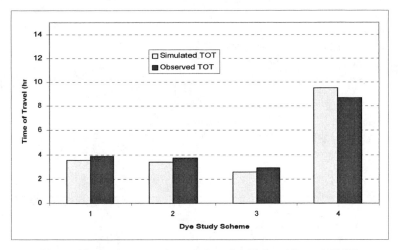

Figure 20. Model-Data Comparison for Time-of-Travel (TOT)

$$\frac{\partial B}{\partial t} = (G - R - P - M)B$$

Where: B is the periphyton/macrophyte biomass per unit area (M/L^2];
         G is the growth rate (T^{-1});
         R is the respiration rate (T^{-1});
         P is the predation rate (T^{-1}); and
         M is the non-predatory mortality rate (T^{-1}).

The effects of nutrients, light, and temperature are taken into account when calculating each term in the above equation, as is done for phytoplankton simulation.

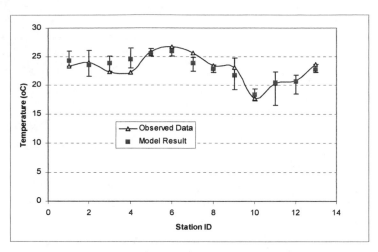

Figure 21. Model-Data Comparison for Water Temperature (bars represent ranges of model-predicted temperatures resulting from diurnal fluctuations simulated)

For application to Wissahickon Creek, minor code modifications were made to the WASP/EUTRO model to fully incorporate oxygen generation and consumption terms by periphyton metabolism in the DO balance equations. In addition, a simplified diurnal simulation module was added to the code to allow for a more accurate representation of DO fluctuation in the receiving water. In this simplified diurnal simulation module, the growth of phytoplankton and periphyton occur during daytime and halt at night. The average radiation intensity was used to govern the algal/periphyton dynamics during daylight hours, and a zero solar radiation intensity was used to restrict algal/periphyton growth during the night. The modified model was capable of simulating time-variable DO with hourly resolution (or higher resolution as necessary), and estimating daily average, minimum, and maximum DO concentrations.

To account for the impact of spatial variability of solar radiation, a new parameter was introduced into the modeling framework scale the solar radiation intensity, thus providing spatial variability. The scaling used for spatial variability was based on canopy cover information provided in the 1998 periphyton survey conducted by PA DEP and ANSP. These initial values were further refined through model calibration.

In the initial WASP/EUTRO model, substrate availability was considered to be spatially-uniform (a global constant). This imposed a severe limitation regarding the capability of the model to realistically simulate the spatial distribution of periphyton impacts. Therefore, the model code was modified to allow the substrate availability to vary segment-wise. Values for substrate availability for each model segment were determined through model calibration.

Three empirical equations are included in the WASP/EUTRO model for prediction of reaeration (O'Connor-Dobbins formula; Churchill formula; and Owens and Gibbs formula) to relate site-specific reaeration coefficients to local streamflow velocity and water depth. The different formulas have various ranges of applicability, and WASP/EUTRO can automatically choose the appropriate formula (by satisfying specific criteria) to calculate the reaeration coefficient. These equations were tested for the Wissahickon Creek system and results showed very high predictions of DO concentrations. Matching observed data would have required unreasonably high SOD values to be used. This condition indicated that these equations, which are essentially empirical, might not be applicable to the Wissahickon Creek watershed. To remedy this problem, a user-defined reaeration equation was incorporated into the WASP/EUTRO framework. This equation was in similar form to the O'Connor-Dobbins formula, however, the coefficients were adjusted during calibration:

$$Ka = A_{coef} V^{Ccoef} H^{Dcoef}$$

Where: Ka is reaeration coefficient (day-1);
         A_{coef}, C_{coef}, and D_{coef} are coefficients subjected to calibration;
         V is water velocity (m/s) and
         H is water depth (m).

The three coefficients A_{coef}, C_{coef}, and D_{coef} were designated as global constants to maintain consistency throughout the system and were obtained through model calibration.

Nine constituents were included as state variables in the water quality model, and boundary conditions were defined for each. Since periphyton is not transportable, the concentration was specified as 0.0 for all twenty boundary conditions. For chlorophyll-*a*, the concentration was specified as 0.0 for the five major point sources, while a background concentration of 0.5 ug/L was specified for the other fifteen boundary conditions. However, the model was found to be insensitive to the chlorophyll-*a* boundary conditions from the headwaters due to the fact that the system is effluent dominant during the modeling period.

Since no data were available for the two organic constituents (organic-N and organic-P), a background concentration of 0.1 mg/L was specified for all twenty boundary conditions. Such low values for organic-N and P concentrations were assumed because Wissahickon Creek is shallow, and the stream velocity is low during the low flow period; this results in quick loss of the organic matter in the water column due to settling and, therefore, has a significant effect on DO concentration through chemical processes within the water column. In addition, the SOD value accounts for the effect of organic matter on DO.

The concentrations of ammonia, nitrate/nitrite, ortho-phosphate, CBODu, and DO for the point sources were specified based on discharger monitoring data and PA DEP data grab samples at the effluent (Table 5). Discharger concentrations of ammonia, nitrate/nitrite, ortho-phosphate, CBODu, and DO in model calibration runs were initially based on composite sample values provided by dischargers. While calibration of key parameters associated with CBODu, ortho-phosphate, nitrate/nitrite, and DO was readily attained, ammonia calibration presented some difficulty. Therefore discharger ammonia concentrations were compared to values obtained from instantaneous grab samples collected by PA DEP. Where differences were observed, PA DEP samples were used.

Table 5. Constituent Concentrations From the Five Major Dischargers

Constituent	Ambler	Upper Gwynedd	Dublin	Abington	North Wales
NH4 (mg/L)	0.03	0.07	0.13	0.06	0.33
NO3 (mg/L)	20.20	14.55	19.32	30.27	15.15
OPO4 (mg/L)	4.68	3.57	2.30	4.63	4.69
CBODu (mg/L)	13.14	5.50	6.00	9.00	6.00
DO (mg/L)	8.62	7.10	7.70	6.74	6.97

The concentrations at boundary conditions for the other 15 locations were determined through an iterative process. First, an initial estimate based on available monitoring data. This estimate was then adjusted within a reasonable range through the calibration process to obtain a refined value.

The nutrient loads from ten minor point sources were configured as "dry" point source loads. "Dry" loads were specified instead of full boundary conditions (complete with flow) because flows from these dischargers were too low (ranging from 0.00007 to 0.08300 cfs) to significantly affect hydrodynamics since the total flow from these minor dischargers only amount to 0.35% of the total flow.

Following model configuration, the model was calibrated for individual parameters through a comparison of model results to observed data gathered during the 2002 low-flow survey period. The main parameters subjected to calibration included algal and periphyton growth rates, respiration rates, and death rates; CBOD decay rate; sediment oxygen demand (SOD); nitrification and denitrification rates; nitrogen and phosphorus recycling rates from dead algae; and carrying capacity of periphyton, substrate availability for periphyton growth, and local solar radiation shading coefficient; and reaeration equation coefficients A_{coef}, C_{coef}, and D_{coef}. The calibration process involved a stepwise adjustment of these parameters, within reasonable and acceptable ranges, until the model adequately reproduced the observed data.

In general, the model reproduced the spatial distribution of water quality very well, particularly in the vicinity of point source discharge locations. In addition, the DO fluctuation, as simulated with the diurnal module, matched the observed data reasonably well. Examples of calibration results are shown in Figures 22 through 24.

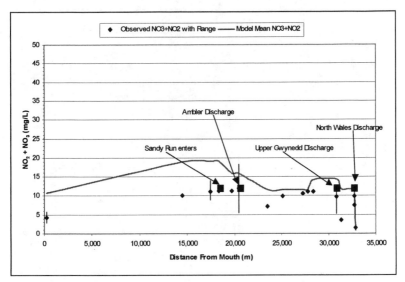

Figure 22. Model Calibration of NO3+NO2 for Wissahickon Creek

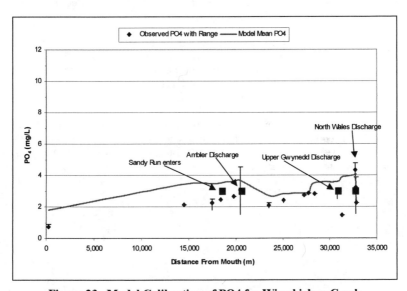

Figure 23. Model Calibration of PO4 for Wissahickon Creek

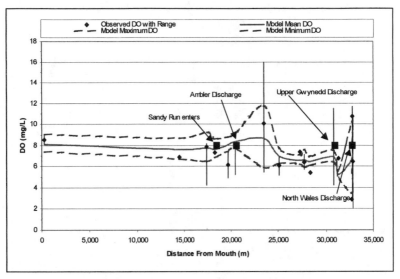

Figure 24. Model Calibration of DO for Wissahickon Creek (bars represent ranges of observed data resulting from diurnal variations; dotted lines are model-predicted diurnal ranges)

Although there were no data indicating the distribution of periphyton in 2002, the model-simulated periphyton was compared with data collected in 1998 to check the capability of the model in simulating the general trend of periphyton. Since the periphyton state variable in the model is defined as mass of carbon, and the collected periphyton data were in terms of chlorophyll-a, a mass conversion operation was necessary before comparisons could be made. Assuming a carbon-to-chlorophyll-a ratio of 30:1, which is similar to the default value for algae (Shanaha and Alam, 2001), and considering a +/- 50 percent range of variation, the simulated periphyton mass was converted to chlorophyll-a (ug/L) and plotted with the surveyed data (Figure 25). As shown, the model captured the general distribution of the periphyton with reasonable accuracy.

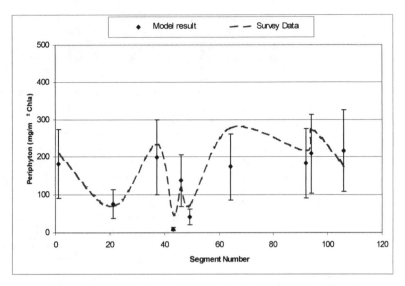

Figure 25. Calibration of Periphyton for Wissahickon Creek

Once calibrated, the model was reconfigured for simulation of critical conditions for TMDL development. The critical condition for DO impairment in Wissahickon Creek is summer low flow. A standard flow often utilized for low-flow, steady-state analysis is the 7Q10 flow, defined as the streamflow that occurs over 7 consecutive days and has a 10-year recurrence interval, or 1 in 10 chance of occurring in any given year. Daily streamflows in the 7Q10 range are general indicators of prevalent drought conditions occurring over large areas. The 7Q10 flow calculated at the mouth (USGS gage 01474000) was 16.26 cfs. This 7Q10 flow is usually extrapolated throughout the upstream and headwaters of a watershed to estimate a basin-wide, steady-state flow for the selected model. For TMDL development and waste load allocations, point sources are modeled at design flows in their respective NPDES permits. However, when all point sources in the Wissahickon Creek basin are at design flows, the combined discharge from point sources is 27.9 cfs, exceeding the 7Q10 flow at the mouth. Since average flows from dischargers are inherently included in the flow budget of Wissahickon Creek through the historical record used for the statistical determination of the 7Q10 flow, this low flow was not determined to define the assimilative capacity of the stream accurately as discharge flows are increased to their design capacity. Therefore, background flows (streamflow without discharge contributions) for Wissahickon Creek were estimated for 7Q10 flow conditions by subtracting average discharge flows recorded during the critical summer period of 2002 (combined flow of 14.9 cfs) from the 7Q10 at the mouth (16.3 cfs). A preliminary estimate of the background flow is thus 1.4 cfs in Wissahickon Creek. After discharges were removed from consideration for 7Q10 flows, the remaining 1.4 cfs flow did not account for flows from Coorson's Quarry (historical average of 12.5 cfs). Under drought conditions, much of the Wissahickon

Creek flow is therefore considered lost to groundwater before reaching the mouth. To accurately simulate the benefits that occur through dilution of Wissahickon Creek streamflows with flows from the quarry, the average of 12.5 cfs was added to Lorraine Run in addition to the background 7Q10 flows distributed throughout the watershed.

The preliminarily estimated flow was distributed to the upper stream of each branch of the Wissahickon Creek basin. These flows were subjected to adjustment for the purpose of maintaining numerical stability in the model. As for the background flow at the downstream portion of Wissahickon Creek, tributary flows were assumed to be the same as those under the drought condition represented by the low flow season of 2002. Once the background 7Q10 flows and quarry flows were configured in the model, permitted discharge flows were added. The resulting total flow at the mouth of Wissahickon Creek for the critical low-flow conditions was 42.52 cfs. Since the flow in the creek is dominated by the discharger flows (accounting for 98.7% of flow in the reaches upstream of the confluence with Lorraine Run), the water quality is insensitive to a moderate magnitude of adjustment or uncertainty in flow estimation. Therefore, the flow estimated in this process can be considered acceptable for further projection analysis. Figure 26 shows the flow distribution for TMDL analysis.

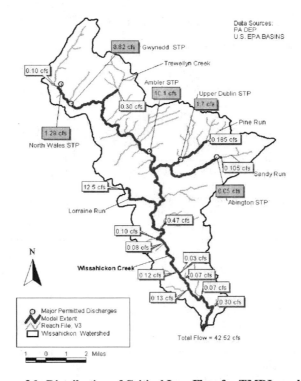

Figure 26. Distribution of Critical Low Flow for TMDL analysis

The calibrated WASP/EUTRO water quality model was updated with critical discharge concentrations from point sources based on either NPDES permit limits for NH4 and CBOD or averages of discharge data from the summer 2002 period for NO3/NO2 and PO4. Note the CBOD permit was in terms of five-day CBOD, which was converted to ultimate CBOD for the model. Using the calibrated decay rate of 0.18/day, the CBODu/CBOD$_5$ ratio was determined to be 1.685 based on the equation:

$$R = \frac{1}{1 - \exp(-5 * Kd)}$$

Where, R is the CBODu to CBOD$_5$ ratio; Kd is the decay rate of CBOD. The discharger's concentration in the reconfigured model is shown in Table 6.

Table 6. Constituent Concentrations From the Five Major Dischargers-Baseline Condition

Constituent	Ambler	Upper Gwynedd	Dublin	Abington	North Wales
NH4 (mg/L)	1.5	1.8	2.5	2.0	2.5
NO3 (mg/L)	20.20	14.55	19.32	30.27	15.15
OPO4 (mg/L)	4.68	3.57	2.30	4.63	4.69
CBODu (mg/L)	16.85	16.85	25.30	16.85	16.85
DO (mg/L)	7.0	7.0	7.0	7.0	7.0

Another important re-configuration of the water quality model was the adjustment of the SOD rate based on the load variation. A linear relationship between SOD reduction and pollutant load reduction was assumed, and this one of the most common approaches for addressing SOD variation following load change (Chapra, 1997). This mechanism was used by Army Corps of Engineers for the Inland Bays Model, and by Hydroqual Inc. for the Appoquinimink Creek model. The algorithm for the linear assumption can be shown as:

$$R_{SOD}(i) = SOD_B(i)[1 + RS(i))]$$

Where, $R_{SOD}(i)$ is the adjustment ratio of SOD rate at segment i;
              $SOD_B(i)$ is the baseline SOD rate at segment i;
              $RS(i)$ is the percent change of load to segment i (positive when load increases; negative when load decreases).

The percent load change was a comprehensive indicator that can be further expressed as:

$$RS(i) = \sum_{k=1}^{n} w_k \, RS_k(i)$$

Where, w_k is the weight factor of constituent k;
              RS_k is the percent load change of constituent k;

Based on this formula, an initial estimate of the adjusted SOD was obtained to account for the load change from the calibrated scenario to the critical discharge scenario. The adjusted SOD at some locations were unreasonably high due to the fact that when the load variations go beyond the applicable range of the linear assumption, the linear equations result in an over-predicted SOD value. Therefore, the values were adjusted to better reflect the characteristics of the system. Since the effluents from the five major dischargers are subject to secondary treatment, the SOD should be at the lower end of the range of 2 to 10 g/ m²/day (Chapra, 1997). Based on this judgment, the maximum SOD value in the creek was determined to be equal or less than 2.1 g/ m²/day downstream of the main dischargers and the values at each segment decrease with distance from the outfall based on the trend identified in the initial estimate.

For TMDL development, model-predicted instream water quality were compared to targets set to provide assurance that designated uses of Wissahickon Creek and tributaries are met or restored. Based on that data and analysis, USEPA determined that the link between nutrient concentrations, DO concentrations, and biological activity in the streams was a necessary component of endpoint determination. This is especially true since biological impacts were a consideration in the original listing of the waterbodies as impaired due to nutrients. Of the components of instream biological activity, only DO has applicable numeric criteria for stream segments of the Wissahickon Creek basin. The standards for DO are based on levels required to support fish populations, with the critical period (period of higher required concentrations) based on supporting the more stringent aquatic life use for trout stocking. This period requires a minimum DO level of 5.0 mg/L and a minimum daily average of 6.0 mg/L to support the aquatic life use for Trout Stocking (TS) from February 15 through July 31. For the remainder of the year, a minimum DO level of 4.0 mg/L and a minimum daily average of 5.0 mg/L are required to support Warm Water Fish (WWF).

The nutrient TMDL endpoints are based on and ensure achievement of both the minimum and minimum daily average DO for the critical periods associated with TS and WWF. However, in analyses of the streams ability to meet these standards, it was necessary to consider all biological processes that are factors in the impairment of the waterbodies. These factors included the link between nutrient levels and biological activity, including effects of periphyton/algae growth and the resulting diurnal variability of DO resulting from biological processes. Through modeling analyses of the Wissahickon Creek and tributaries, instream DO concentration was predicted to be highly sensitive to those parameters directly related to periphyton growth and respiration. In addition, shading or exposure to direct sunlight is also a fairly sensitive factor impacting DO concentration. Other relatively sensitive factors include sediment oxygen demand and stream reaeration.

4.2   Siltation Modeling Approach

To develop a siltation TMDL for the impaired reaches in the basin, a "reference watershed approach" was utilized. Pennsylvania does not currently have numeric criteria for siltation. Therefore, a reference watershed approach was used to establish numeric endpoints for sediment in Wissahickon Creek. The reference watershed approach is used to estimate the necessary load reduction of sediment that would be needed to restore a healthy aquatic community and allow the streams in the watershed to achieve their designated uses. This approach is based on determining the current loading rates for the pollutants of interest from a selected unimpaired watershed that has similar physical characteristics (i.e., landuse, soils, size, geology) to those of the impaired watershed. The approach pairs two watersheds, one attaining its uses and one that is impaired based on biological assessment. The objective of this process is to reduce the loading rate of sediment (or other pollutant) in the impaired stream segment to a level equivalent to or slightly lower than the loading rate in the unimpaired reference stream segment. Achieving the sediment loadings set forth in the TMDLs will ensure that the designated aquatic life of the impaired stream is achieved.

The reference watershed selection process is based on a comparison of key watershed and stream characteristics. The goal of the process is to select one or several similar, unimpaired reference watersheds that can be used to develop TMDL endpoints. Reference watershed selection was based on a desktop screening of nearby non-impaired watersheds with characteristics similar to those of the Wissahickon Creek watershed using several GIS coverages. The GIS coverages included the USGS watershed coverage, the state water plan boundaries, the satellite image-derived land cover grid (MRLC), stream reach coverage, Pennsylvania's 305(b) assessed streams database, the STATSGO soils database, and geological coverages.

Based on the aforementioned desktop GIS search for a reference watershed, the Ironworks Creek watershed, located in Bucks and Montgomery counties, was used to establish reference conditions and TMDL endpoints for the Wissahickon Creek watershed. The reference watershed was chosen based on the fact that it was an urban watershed that was not impaired by siltation and had similar physical characteristics to the Wissahickon Creek watershed (i.e., watershed size, landuse/cover, soils, geology, ecoregion). Table 7 presents the characteristics of both the Wissahickon Creek and Ironworks Creek watersheds. Figure 27 presents the location of the Ironworks Creek watershed relative to Wissahickon Creek.

Table 7. Impaired and Reference Watershed Comparison

	Wissahickon Creek	Ironworks Creek
Watershed Type	*Impaired Watershed*	*Reference Watershed*
Watershed Size (acres)	40,928	11,114
Geologic Province	Piedmont	Piedmont
Dominant Rock Types	Sandstone/Metamorphic-Igneous/Shale/Carbonate	Sandstone/Metamorphic-Igneous
Dominant Soils	C & B	C & B
Ecoregions	Triassic Lowlands Piedmont Uplands Piedmont Limestone Dolomite Lowlands	Triassic Lowlands Piedmont Uplands
Percent Slope of Watershed	0.25%	0.63%
Point Sources	14	0
Percent Urban	43%	44%
Percent Forested	40%	31%
Landuse Types:	% Landuse	% Landuse
Low Intensity Development	34.1%	39.8%
High Intensity Development	8.5%	4.2%
Hay/Pasture	7.1%	11.7%
Cropland	8.9%	10.9%
Conifer Forest	2.4%	1.8%
Mixed Forest	10.2%	10.3%
Deciduous Forest	28.0%	19.6%
Quarry	0.3%	0.0%
Coal Mine	0.02%	0.0%
Transitional	0.4%	0.1%

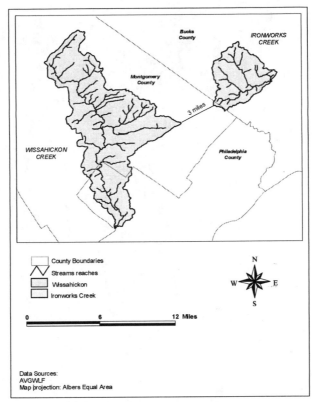

Figure 27. Location of Reference Watershed for Siltation TMDL Development

Wissahickon Creek is a much larger watershed (40,928 acres) than Ironworks Creek (11,114 acres), therefore, Wissahickon Creek was delineated into five smaller subwatersheds that could easily be compared to Ironworks Creek. Ironworks Creek was subsequently re-delineated to appropriately match each of the five subwatersheds in the Wissahickon Creek watershed (Figure 28).

To equate target and reference watershed areas for TMDL development, the total area for the reference watershed was adjusted to be equal to the area of its paired target watershed, after calibration of hydrology to observed data. To accomplish this, land use areas (in the reference watershed) were proportionally adjusted based on the percent land use distribution. As a result, the total watershed area for Ironworks Creek was adjusted to be equal to the five modeled subwatersheds in Wissahickon Creek, respectively.

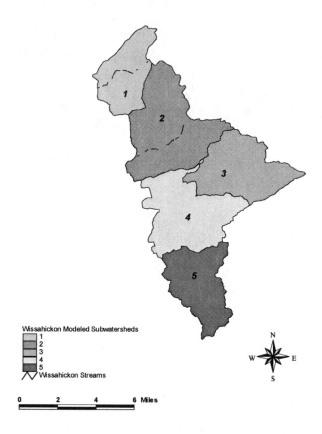

Wissahickon Modeled Subwatersheds
1
2
3
4
5
Wissahickon Streams

0   2   4   6 Miles

Figure 28. Five Subwatersheds of Wissahickon Creek

Once the impaired and reference watersheds were matched, a watershed model was used to simulate the sediment loads from different sources. The modeling framework used in this study consisted of a modified application of the Generalized Watershed Loading Function (GWLF) watershed model (Haith and Shoemaker, 1987), including a special module for simulation of streambank erosion. The ArcView Version of the Generalized Watershed Loading Function (AVGWLF), developed by the Environmental Resources Research Institute of the Pennsylvania State University (Evans et al. 2001), was utilized for development of GWLF model input and estimation of sediment loadings from overland runoff. Using hydrology input parameters established by the AVGWLF model, BasinSim (Dai et al., 2000) was used to run GWLF with model output specially formatted for a separate Streambank Erosion Simulation Module. Loadings from streambank erosion were estimated with this separate module using daily flows predicted by GWLF, site-specific information, and process-based algorithms.

GWLF is an aggregated distributed/lumped parameter watershed model. For surface loading, it is distributed in the sense that it allows multiple land use/cover scenarios. Each area is assumed to be homogenous with respect to various attributes considered by the model. Additionally, the model does not spatially distribute the source areas, but aggregates the loads from each area into a watershed total. In other words, there is no spatial routing. Daily water balances are computed for an unsaturated zone as well as for a saturated subsurface zone, where infiltration is computed as the difference between precipitation and snowmelt minus surface runoff plus evapotranspiration.

GWLF models surface runoff using the Soil Conservation Service Curve Number (SCS-CN) approach with daily weather (temperature and precipitation) inputs. Erosion and sediment yield are estimated using monthly erosion calculations based on the Universal Soil Loss Equation (USLE) algorithm (with monthly rainfall-runoff coefficients) and a monthly composite of KLSCP values for each source area (e.g., land cover/soil type combination). The KLSCP factors are variables used in the calculations to depict changes in soil loss/erosion (K), the length/slope factor (LS), the vegetation cover factor (C), and the conservation practices factor (P). A sediment delivery ratio based on watershed size and a transport capacity based on average daily runoff are applied to the calculated erosion to determine sediment yield for each source area. Evapotranspiration is determined using daily weather data and a cover factor dependent on land use/cover type. Finally, a water balance is performed daily using supplied or computed precipitation, snowmelt, initial unsaturated zone storage, maximum available zone storage, and evapotranspiration values. All of the equations used by the model can be found in the original GWLF paper (Haith and Shoemaker, 1987) and GWLF User's Manual (Haith et. al, 1992).

The streambank erosion simulation module employed the algorithm used in the Annualized Agricultural Nonpoint Source Model (AnnAGNPS) model (Theurer and Bingner, 2000). Sediment transport/routing and streambank erosion simulation were performed using three particle size classes (clay, silt, and sand), with a simulation time-step of one hour. For each subwatershed channel segment, the incoming sediment load is the total of local sources plus the loading from upstream subwatersheds. If the incoming sediment load was greater than the sediment transport capacity specific to the physical features and the magnitude of flow of that segment, then the sediment deposition algorithm was used. If the incoming sediment is less than or equal to the sediment transport capacity, the sediment discharge at the outlet of the segment will be equal to the sediment transport capacity for an erodible channel. Since the sediment transport capacity is specific to the magnitude of flow, the capacity for each particle size was calculated for each increment of the streamflow hydrograph. The erodibility of a channel was reflected by the sediment availability factor for the three particle sizes. These factors were assigned based on site-specific information regarding bank stability and vegetation cover conditions, and were further calibrated where monitoring data was available.

Local rainfall and temperature data were used to simulate flow conditions in modeled watersheds. Hourly precipitation and daily temperature data were obtained from local

National Climatic Data Center (NCDC) weather stations and other sources. Daily maximum and minimum temperature values were converted to daily averages for modeling purposes. The period of record selected for model runs (April 1, 1993 through March 31, 2001) was based on the availability of recent weather data and corresponding streamflow records. The weather data collected at the NCDC station Palm 3 SE located approximately 15 miles northwest of the Wissahickon Creek watershed were used to construct the weather file used in all watershed simulations (both impaired and reference).

Sediment point sources were not included in the GWLF model because GWLF is set up to include nutrient point sources, but not sediment point sources. There are 14 point sources of sediment in the Wissahickon Creek watershed. The sediment loads (in lbs/yr) from these point sources were calculated outside of the model based on their permitted flow and TSS concentration. The sediment delivery ratio for the watershed in which each point source was located was applied to the total sediment load from that point source to determine the resulting sediment load at the mouth of the watershed after transport losses.

Daily streamflow data are needed to calibrate watershed hydrology parameters in the GWLF model. There is a continuous USGS flow gage at the mouth of Wissahickon Creek (USGS 01474000) that has flow data from October 1, 1965 through September 30, 2001. There is no flow gage in the reference watershed of Ironworks Creek, so hydrology was calibrated using data collected at the nearby Little Neshaminy Creek watershed, which is similar in size as well as other characteristics (i.e., soils, geology, landuse) to Ironworks Creek. The Little Neshaminy gage (USGS 01464907 Little Neshaminy Creek @ Valley Rd near Neshaminy, PA) has flow data from November 25, 1998 through September 30, 2001.

Using the input files created in AVGWLF, the model predicted overall water balances in impaired and reference watersheds. In general, an R^2 value greater than 0.7 indicates a strong, positive correlation between simulated and observed data. The R^2 value for the Wissahickon Creek and Ironworks Creek hydrology calibrations were 0.76 and 0.74, respectively. These results indicate a good correlation between simulated and observed results for these watersheds. Hydrology calibration results for Wissahickon Creek and the reference watershed are presented in Figures 29 and 30.

To provide the necessary input files for the streambank erosion model, the AVGWLF model parameters and input files were reformatted for input into BasinSim. The BasinSim watershed simulation program is a windows-based modeling system that facilitates the development of model input data to run GWLF model, and provides additional post-processing functionality (Dai et al., 2000). BasinSim has been modified to generate the required input files to run the stream routing and bank erosion simulation module. As a result, the five Wissahickon Creek subwatersheds were configured and routed separately in BasinSim.

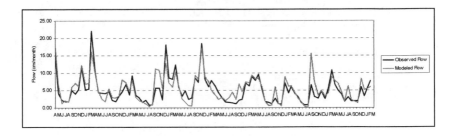

Figure 29. Hydrology Calibration at USGS Gage 01474000 (Wissahickon Creek at Mouth); April 1993 through March 2001

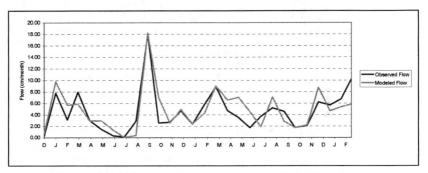

Figure 30. Hydrology Calibration for Ironworks Creek Using the Reference Gage at Little Neshaminy Creek (USGS 01464907); December 1998 through March 2001

To simulate routing, modifications were made to BasinSim to provide the functionality and hydraulic methods required. The upstream inflow hydrograph was routed through the receiving channel to calculate the downstream outflow hydrograph. The Muskingum-Cunge method was used for routing calculations (Maidment, 1993; Viessman et al., 1996). The Muskingum-Cunge method uses physical characteristics (e.g., channel length, slope, kinematic wave celerity, and a characteristic unit width discharge) to compute hydrograph routing through the channel.

Following configuration of the BasinSim model with multiple subwatersheds and routing procedures, this application was verified against results from the AVGWLF modeling approach where the Wissahickon Creek watershed was calibrated with observed data (Figure 29). Results showed consistency between previously modeled flow and the new routed flow procedures (Figure 31).

The BasinSim model results were used to drive the Streambank Erosion Simulation Module, considering both streamflow routing and sediment transport. As the routing and streambank erosion simulation uses hourly or smaller time step, the daily GWLF flow was extrapolated to a triangular hydrograph at an hourly increment by using extended TR-55 procedures consistent with methodologies utilized by AnnAGNPS (Theurer and Bingner, 2000; Bingner et al., 2001).

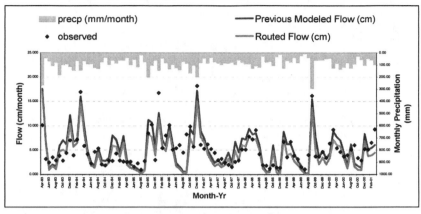

Figure 31. BasinSim Hydrology Verification at USGS 01474000 (Wissahickon Creek at Mouth); April 1993 through March 2001

Water quality calibration of the watershed model to observed data considered site-specific factors associated with both overland runoff and streambank erosion. As a result, both the AVGWLF model and Streambank Erosion Simulation Module input parameters were subject to modifications with guidance provided by field observations. Site-specific input data for the GWLF model were provided by the AVGWLF interface. Site-specific information for the Streambank Erosion Simulation Module were obtained from field surveys performed by the Pennsylvania Department of Environmental Protection (PA DEP), and GIS data included in AVGWLF. Due to the empirical formulation of GWLF and the availability of site-specific, regionally verified input data from AVGWLF and PA DEP field surveys, partial calibration was only determined necessary where data was available.

Water quality observations at the same location as USGS gage 01474000 at the mouth of Wissahickon Creek were available (as a concentration) to compare to model output, however, sediment loading rates are predicted by GWLF as monthly loads. The average daily streamflow and monthly TSS concentrations in mg/L were used to determine an estimated monthly sediment load based on linear regression. Since historical water quality observations were on a monthly basis, this data provided limited information on wet weather loads since much of the data were collected during dry periods. Therefore, a linear regression analysis was performed to determine TSS concentrations as a function of flow at the mouth of Wissahickon Creek. Based on the comparison of the model output to observed TSS values (determined using linear regression) for the period of January 1994 through December 2000, the Wissahickon Creek watershed's C (vegetation cover) and P (conservation practices) input values for GWLF were adjusted to reflect the high sediment loads observed in the watershed. Observed water quality data were not available for comparison to reference watershed output, therefore the default sediment parameters selected during GWLF setup were used. Based on habitat assessments provided by PADEP for stream segments in the Wissahickon Creek watershed as well as the Ironworks Creek watershed, the Wissahickon Creek watershed had poorer habitat conditions than Ironworks creek, which supports the increased C and P values used in modeling the Wissahickon Creek watershed.

Results of field surveys performed in 1998 were used to determine conditions of streambanks and the degree of vegetative protection present for estimation of streambank erosion. The field survey recorded the condition of banks and respective vegetative protection on a scale of 20 to 0, with higher numbers representing more stable banks and more vegetative cover. Using the average of the bank condition score and vegetative cover score, relative sediment availability factors were estimated for the channel of each modeled subwatershed. The magnitudes of sediment availability weighting factors for clay, silt and sand were further calibrated using available sediment monitoring data collected at the mouth of Wissahickon Creek.

Results of calibration of the watershed model to monthly sediment load (predicted through regression analysis of TSS data and streamflow) are shown in Figure 32. Model results are shown with and without streambank erosion to illustrate the marginal contributions of each source of siltation.

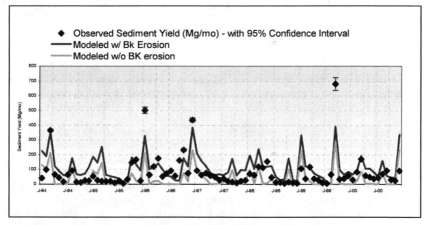

Figure 32. Calibration of the Watershed Model to Water Quality Data Collected at the Mouth of Wissahickon Creek

Through modeling of the reference watershed, reference loads were estimated for each of the five model subwatersheds of Wissahickon Creek using loads estimated from the reference watershed. These reference loads determined the TMDL for each model subwatershed. Following analysis of model results, land-use-specific, unit-area loads (UALs) were calculated for each subwatershed of Wissahickon Creek and the reference watershed. Using these UALs as guides, specific land uses within each subwatershed were subject to reductions in siltation loads to meet reference conditions (TMDL). Resulting UALs for each subwatershed to meet TMDLs are reported in Table 8. Using the land use distribution within each municipal boundary, wasteload allocations were calculated for overland siltation loads.

Table 8. Unit-Area Loading Rates for Siltation by Land Use

Subwatershed	Unit Area Loading Rate (lbs/acre/yr)				
	1	2	3	4	5
Low-Intensity Residential	164.62	173.45	180.50	258.93	420.17
High-Intensity Residential/Urban	139.41	129.28	137.11	106.22	278.76
Hay/Pasture	51.60	48.02	76.84	42.54	108.17
Row Crops	464.28	301.79	306.60	254.55	623.34
Coniferous Forest	3.13	2.74	4.94	5.74	8.82
Mixed Forest	3.99	3.93	5.67	4.81	9.43
Deiduous Forest	5.43	4.58	7.00	8.69	32.00
Quarry	0.00	0.00	0.00	619.45	0.00
Coal Mines	0.00	0.00	0.00	352.72	0.00
Transitional	0.00	0.00	526.14	751.42	12931.69

5.0   TMDL Development

Separate methodologies were utilized for determination of nutrient and siltation TMDLs. Each selected methodology considers specific impacts and conditions determined necessary for accurate source representation and system response.

5.1   Nutrient TMDL

The modeling system for nutrient TMDL development was first configured and calibrated for low-flow conditions observed in summer 2002 using data collected by USGS, PA DEP, and major dischargers in the watershed. Once calibrated, the modeling system was configured for 7Q10 flow conditions to assess "baseline" conditions in the stream. To achieve water quality endpoints in the stream segments, multiple scenarios were modeled to account for varying discharge concentrations and conditions. Optimal results were reached that met instream water quality endpoints with minimal impact to stakeholders. However, reductions were required from dischargers so that these endpoints could be met.

Separate TMDLs were established for each individual stream segment listed on Pennsylvania's 303(d) list. Each TMDL consists of a point source wasteload allocation (WLA), a nonpoint source load allocation (LA), and a margin of safety (MOS). These TMDLs identify the sources of pollutants that cause or contribute to the impairment of the DO criteria and allocate appropriate loadings to the various sources. Given the scientific knowledge available, and utilizing the model processes that describe the interrelationship of nutrients, carbonaceous oxygen demand (CBOD), sediment oxygen demand (SOD), and their impact on DO, EPA determined that the appropriate pollutants for these TMDLs, LAs and WLAs are ammonia, nitrate and nitrite, ortho phosphate, and CBOD.

The equation used for TMDLs and allocations to sources is:

$$TMDL = WLA + LA + MOS$$

The WLA portion of this equation is the total loading assigned to point sources. The LA portion is the loading assigned to nonpoint sources. The MOS is the portion of loading reserved to account for any uncertainty in the data and the computational methodology used for the analysis. For this study, the MOS is assumed implicit through conservative assumptions and the steady-state modeling approach of low flow conditions.

Federal regulations (40 CFR 130.7) require TMDLs to include individual WLAs for each point source. Of the twenty-three National Pollution Discharge Elimination System (NPDES) permitted dischargers, only five facilities are likely to require reductions to their respective NPDES permit limits for the pollutants considered.

Using the model described above, EPA made these allocations by reducing CBOD, ammonia nitrogen, nitrate and nitrite, and ortho phosphate loads from NPDES point sources until daily average and minimum daily DO criteria were satisfied. WLAs for each point source were determined on a case-by-case basis, with most reductions determined by local improvements downstream from the point of discharge. Where dischargers were in close proximity, sensitivity analyses were performed to ensure that appropriate sources received reductions.

At the request of stakeholders, effluent water quality from Ambler Borough (PA0026603), Upper Gwynedd Township (PA0023256), Abington Township (PA0026867), Upper Dublin Township (PA0029441), and North Wales Borough (PA0022586) were modeled assuming DO concentrations of 7.0 mg/L, which is higher than levels presently specified by NPDES permits for each discharger. This was justified because higher DO concentrations are generally provided by these dischargers. One of the assumptions for each of these WLAs is that the effluent DO concentration will be raised to 7.0 mg/l. These WLAs therefore require not only that the pollutant specific limits be consistent, but also that the facilities achieve a DO effluent concentration of no less than 7.0 mg/l.

Nonpoint source loads within the Wissahickon Creek basin were based on low-flow samples collected by PA DEP in summer 2002. Water quality samples were taken at upstream locations and select tributaries to estimate background loads. These loads were included in the calculations of TMDLs. However, no load reductions were determined necessary for background loads. As a result, impaired stream segments without associated major point sources required no reductions for either WLAs nor LAs because water quality data did not suggest that such reductions were warranted. However, to address the impairments in these stream segments, implementation measures were recommended to address non-source related factors that can result in biological improvements (e.g., increased tree canopy to improve shading).

TMDLs were developed for each of the seasonal water quality criteria for DO applicable to the Wissahickon Creek basin and include: (1) Trout Stocking (TS) from February 15 to July 31, and (2) Warm Water Fishes (WWF) for the remainder of the year. For each stream segment in the Wissahickon Creek basin included in Pennsylvania's 303(d) list due to nutrients (Figure 33), separate TMDLs, WLAs, and LAs were determined for both TS and WWF periods, respectively. Total loads were determined for CBOD5, ammonia nitrogen, nitrate-nitrite nitrogen, and ortho phosphate. For each of the five major dischargers, WLAs are listed in Tables 9 and 10 for TS and WWF DO criteria, respectively. WLAs are specific to the summer period. For the remainder of the year, implementation of WLAs require seasonal adjustments following PA DEP procedures (PA DEP, 1997).

Table 9. WLAs for Five Major Dischargers in the Wissahickon Creek Watershed - Trout Stocking (April 1/May 1 to July 31)

Name	NPDES	CBOD5 (mg/L)	NH3-N (mg/L)	NO3+NO2-N (mg/L)	Ortho PO4-P (mg/L)	TMDL Percent Reduction CBOD5[A]	NH3-N[A]	NO3+NO2-N[B]	Ortho PO4-P[B]
North Wales Boro	PA0022586	3.00	0.50	15.16	1.41	70.0%	80.0%	0.0%	70.0%
Upper Gwynedd Township	PA0023256	5.00	0.74	20.08	1.82	50.0%	59.0%	0.0%	49.0%
Ambler Boro	PA0026603	10.00	1.50	30.52	4.68	0.0%	0.0%	0.0%	0.0%
Abington Township	PA0026867	7.50	0.72	30.27	1.85	25.0%	64.0%	0.0%	60.0%
Upper Dublin Township	PA0029441	12.77	2.25	36.71	1.45	14.9%	10.0%	0.0%	36.9%

A - Calculated from NPDES permit limit
B - Calculated from average of summer 2002 monitoring. If allocations exceeded average of 2002, 0.0 % is reported.

Table 10. WLAs for five major dischargers in the Wissahickon Creek watershed - Warm Water Fishes (August 1 to October 31)

Name	NPDES	CBOD5 (mg/L)	NH3-N (mg/L)	NO3+NO2-N (mg/L)	Ortho PO4-P (mg/L)	TMDL Percent Reduction CBOD5[A]	NH3-N[A]	NO3+NO2-N[B]	Ortho PO4-P[B]
North Wales Boro	PA0022586	5.90	1.37	21.22	2.40	41.0%	45.0%	0.0%	49.0%
Upper Gwynedd Township	PA0023256	8.50	1.62	19.05	3.22	15.0%	10.0%	0.0%	9.9%
Ambler Boro	PA0026603	10.00	1.50	30.31	4.68	0.0%	0.0%	0.0%	0.0%
Abington Township	PA0026867	10.00	2.00	30.27	4.63	0.0%	0.0%	0.0%	0.0%
Upper Dublin Township	PA0029441	15.00	2.50	32.85	2.30	0.0%	0.0%	0.0%	0.0%

A - Calculated from NPDES permit limit
B - Calculated from average of summer 2002 monitoring. If allocations exceeded average of 2002, 0.0 % is reported.

5.2 Siltation TMDL

For this study, separate approaches for TMDL calculation were used for determination of WLAs associated with overland runoff and streambank erosion, with different MOS assumptions for each. For overland runoff, an explicit MOS of 10% was assumed to ensure protection of the stream segments. For streambank erosion, due to the conservative assumptions regarding allocation of loads throughout the watershed, an implicit MOS was assumed (i.e., no numeric MOS for TMDL calculation).

Of the 13 NPDES dischargers permitted to discharge specific amount of sediment (measured as TSS), none required reductions to their NPDES permit limits (e.g., treated sewage effluents). Based on available discharge monitoring reports (DMR) the average discharge of sediment from such facilities in the watershed was usually well below the permitted TSS concentration.

Figure 33. Stream Segments and Respective Watersheds Listed for Siltation in the Wissahickon Creek Watershed

Stormwater permits typically do not have numeric limits for sediment. EPA's stormwater permitting regulations require municipalities to obtain permit coverage for all stormwater discharges from separate storm sewer systems (MS4s). For these discharges, WLAs were determined using land-use-specific, unit-area loads determined in modeling analysis for specific regions of the Wissahickon Creek basin, as well as the streambank erosion within each municipality. Sediment loads were estimated for each of the five modeled subwatersheds and then distributed among municipalities as MS4 stormwater permit loads (WLAs) for each individual 303(d) listed watershed. Distribution of loads was accomplished within the five subwatersheds for all 303(d) listed watersheds and municipalities based on the corresponding unit-area loading (lbs/acre/year) for overland runoff and streambank erosion determined though modeling analysis.

Since the Wissahickon Creek watershed was divided into five smaller subwatersheds to better match the reference watershed size, sediment allocations began at the top of the watershed (i.e., subwatershed 1) and continued downstream to the mouth of the watershed (i.e., subwatershed 5). After sediment reductions sufficient to achieve and maintain water quality standards were made to the first subwatershed (subwatershed 1) based on the sediment load in the reference watershed, the resulting reduced sediment load was added to the next downstream subwatershed (subwatershed 2) to represent the in-stream sediment load coming from upstream. The sediment load coming from subwatershed 1 was subjected to the sediment delivery ratio (SDR) for subwatershed 2 to account for natural losses. The same upstream load was also added to the reference watershed to account for loading from upstream. The total sediment load in the subwatershed was then compared to the reference watershed sediment load so that reductions could be made. This process continued downstream to the mouth of the Wissahickon Creek watershed. As the reduced sediment loads from upstream Wissahickon Creek were added to the downstream subwatersheds, no further reductions were made to the upstream loads since they were already meeting the appropriate reference watershed sediment target.

For each stream segment in the Wissahickon Creek basin included on Pennsylvania's 303(d) list due to siltation, separate TMDLs, WLAs, and LAs were determined. Total sediment loads from land uses within the Wissahickon Creek watershed were based on unit-area loadings for each landuse. The streambank erosion sediment load was distributed to each of the listed segments in the appropriate watershed based on the drainage area of each listed segment (i.e., if a particular listed watershed made up 12 percent of the larger modeled subwatershed, it received 12 percent of the streambank erosion load).

Each municipal source (MS4 stormwater permit) received a WLA based on the sediment loading from land uses and streambank erosion within the municipal boundaries. The only load reductions to meet TMDLs were assigned to WLAs associated with stormwater permits. No load reductions were required of WLA or LAs associated with WWTP discharges or background loads, respectively. The individual WLAs for each municipal area are presented as a total for each township in Table 11.

6.0   Implementation

Development of TMDLs is only the beginning of the process for stream restoration and watershed management. Load allocations to point and nonpoint sources serve as targets for improvement, but success is determined by the level of effort put forth in making sure that those goals are achieved. Load reductions proposed by nutrient and siltation TMDLs require specific watershed management measures to ensure successful implementation.

6.1   Nutrient TMDL Implementation

Implementation of best management practices (BMPs) in conjunction with wasteload reductions from point sources should eventually achieve the loading reduction goals established in the TMDLs. Further "ground truthing" should be performed in order to assess both the extent of existing BMPs, and to determine the most cost-effective and environmentally protective combination of BMPs required for meeting the required nutrient reductions.

Table 11. Summary of Sediment Wasteload Allocations for Streambank Erosion and Overland Load by Municipality (MS4)

Municipality	Existing Load from Streambank Erosion (lbs/yr)	Streambank Erosion WLA (lbs/yr)	Percent Reduction for Streambank Erosion	Existing Overland Load (lbs/yr)	Overland Load WLA (lbs/yr)	Percent Reduction for Overland Load (lbs/yr)	TOTAL WLA (lbs/yr)
Abington	121,604	41,117	0.66	362,539	87,797	0.76	128,913
Ambler	17,974	9,347	0.48	75,009	32,843	0.56	42,190
Cheltenham	1,758	1,512	0.14	20,549	4,449	0.78	5,961
Horsham	2,611	1,267	0.51	5,764	2,289	0.60	3,556
Lansdale	10,032	5,217	0.48	60,296	47,116	0.22	52,332
Lower Gwynedd	168,245	87,488	0.48	575,511	349,873	0.39	437,360
Montgomery	25,443	13,231	0.48	135,550	97,898	0.28	111,128
North Wales	8,414	4,376	0.48	50,071	37,956	0.24	42,332
Philadelphia	133,827	115,091	0.14	1,413,863	265,770	0.81	380,861
Springfield	51,241	38,361	0.25	700,517	151,804	0.78	190,165
Upper Dublin	350,903	131,126	0.63	906,099	333,482	0.63	464,608
Upper Gwynedd	73,016	37,969	0.48	695,875	512,616	0.26	550,584
Upper Moreland	1,108	367	0.67	1,303	495	0.62	862
Whitemarsh	79,222	51,035	0.36	479,267	188,498	0.61	239,532
Whitpain	105,138	55,148	0.48	357,776	236,125	0.34	291,273
Worcester	1,423	740	0.48	10,645	9,610	0.10	10,350

For stream segments of Trewellyn Creek (971217-1145-ACE), Lorraine Run (971215-1000-ACE), and headwaters of Pine Run (971215-1300-ACE), no reductions from point sources were necessary because either none were present or data was not available to suggest that DO criteria were not being met. Data was simply used for model calibration or verification that there was an impairment. For these segments, it was assumed that biological conditions in the stream are most likely caused by environmental factors that can be remedied through proper management techniques, rather than a result of load reductions in the stream. Specific BMPs are suggested by EPA to provide assurance that biological improvements are provided for these stream segments. Poor biological conditions are considered to be controlled by two primary factors for these segments: (1) extremely shallow conditions in the stream caused by lack of baseflow, and (2) lack of sufficient shading to naturally reduce the biological activity stimulated by higher water temperatures resulting from exposure to direct sunlight. To provide additional baseflow for the low-flow period, BMPs are recommended that encourage infiltration through either stormwater retention or stream buffer zones. Such management practices would also address those stream segments of the Wissahickon Creek basin included on the 303(d) list as a result of impairments associated with water/flow variability. To increase shading, EPA recommends that additional tree canopy be provided along the stream banks.

Several other stream segments will benefit from similar BMPs in conjunction with upstream waste load reductions. Additional tree canopy can potentially reduce biological activity causing diurnal variability of DO concentrations resulting in violations of water quality standards. In addition, BMPs that seek to increase baseflow can result in additional assimilative capacity of the stream for point source discharges.

The nutrient TMDL and WLAs are contingent on the assumption that NPDES permits for the five significant municipal facilities increase the effluent DO concentrations to 7.0 mg/L as a daily minimum. To provide flexibility in implementation, equally protective TMDLs and WLAs were determined for several scenarios: (1) all major discharges with DO levels at 6.0 mg/L (includes required increases from Ambler Borough and Abington Township), (2) all major dischargers with DO levels at 7.0 mg/L, (3) all major dischargers with DO levels at 7.5 mg/L, (4) all major dischargers with DO levels at 7.75 mg/L, and (5) all major dischargers with DO levels at 8.0 mg/L. These scenarios will be used as guidance for PA DEP in reissuing NPDES permits so that TMDLs are met.

This TMDL considered the implementation of seasonal limits. Tables 9 and 10 present the recommended allocations for two seasonal periods when this TMDL is applicable. In addition, PA DEP has established a seasonal effluent limitations strategy for permitting point sources. This strategy is documented in DEPs policy "Determining Water Quality-based Effluent Limits," December 9, 1997. This strategy establishes a set of seasonal "multipliers" for various conventional and non-conventional pollutants. Table 12 presents these multipliers for the pollutants covered under this TMDL. Note that PA DEP has not included a multiplier for DO or nitrate-nitrite nitrogen (NO_2-NO_3-N). For this TMDL, EPA has assumed that the multiplier for NO_2-NO_3-N is the same as that for phosphorus.

Table 12. Seasonal Multipliers Based on PA DEP Seasonal Effluent Limitations Strategy

Parameter	Seasonal Time Period	Winter Limit Multiplier
BOD	Nov 1 - Apr 30	2.0
Phosphorus	Nov 1 - Mar 31	2.0
Ammonia	Nov 1 - Apr 30	3.0

Based on these multipliers and seasonal time periods for the pollutants of concern, winter seasonal limits were determined. Note that this TMDL did not include water quality modeling for the winter period and the winter limits are based solely on PA DEP's strategy. Modifications to these winter limits can be made with no impact on this TMDL. Table 13 presents the winter limits for the five major municipal facilities considered in this TMDL. These winter limits are based on two separate periods. Since the TS standard applies from mid-February through June, the winter multipliers for the period mid-February to May 1 for BOD and mid-February through April 1 for phosphorus and nitrate-nitrite-N were applied to the allocations determined for the low flow TS period. The WWF standard applies from July through mid-February so the winter multipliers for the period from November to mid-February for BOD and November through mid-February for phosphorus and nitrate-nitrite-N were applied to the allocations determined for the low flow WWF period.

Table 13. Seasonal Limits Based on PA DEP's Strategy (mg/L)

Pollutants	North Wales (mg/L)	Upper Gwynedd (mg/L)	Ambler (mg/L)	Abington (mg/L)	Upper Dublin (mg/L)
BOD (Nov 1 - Feb 15)	11.8	17	20	20	30
BOD (Feb 15 - April 30)	6.0	10.0	20.0	15.0	25.5
Ortho P04-P (Nov 1 - Feb 15)	4.8	6.4	9.2	9.3	4.6
Ortho P04-P (Feb 15 - March 31)	2.8	3.6	9.2	3.7	2.95
NO2-NO3-N (Nov 1 - Feb 15)	No Limit	No Limit	No Limit	No Limit	No Limit
NO2-NO3-N (Feb 15 - March 31)	No Limit	No Limit	No Limit	No Limit	No Limit
NH3-N (Nov 1 - Feb 15)	4.1	4.9	4.5	6	7.5
NH3-N (Feb 15 - April 30)	1.5	2.22	4.5	2.16	6.75

6.2   Siltation TMDL Implementation

The 64-square-mile Wissahickon Watershed comprises a variety of land uses from urban to suburban to forest and parkland. The mainstem of the Creek traverses southeasterly for 24 miles through 16 Townships and several boroughs, from the headwaters in Lansdale to the mouth at the Schuylkill River in Philadelphia's Fairmount Park. The banks and surrounding land around the Wissahickon Creek vary as the Creek travels through each township and borough. The specific methods used to address high pollutant load reductions will vary with the land use along the particular segment of Creek. The methods used will also vary depending on the particular source of the pollutant load whether it is stream bank erosion from high flow conditions or overland flow which carries the pollutants from surrounding land.

This TMDL allocates the siltation contributed by both streambank erosion and overland stormwater runoff as WLAs. These WSAs are characterized as such due to the fact that the Wissahickon Creek watershed is in an urbanized area that is regulated by the NPDES Program for MS4 discharge of stormwater. While the loads can be grossly attributed to the MS4s as municipal point sources, the actual contribution of sediment may in some areas be due to "nonpoint sources" as well, including agricultural activities, forested lands, industrial activities, and other sources regulated and unregulated through the stormwater program.

The relative contribution of siltation by both sources varies throughout the watershed according to the distribution of land uses between urban and non-urban, and the amount of impervious cover in the watershed. Streambank erosion is the most significant contributor. Therefore, reductions in the siltation entrained in overland flow must be accompanied by substantial reductions in the volume of water delivered to the stream in order to achieve the water quality objectives of the TMDL. Efforts must also be taken to control future potential sources of sediment and stormwater as new construction and redevelopment occurs. Because of the complexity of the problem and the potential solutions, an adaptive approach will be needed to achieve the TMDLs.

Both regulatory and nonregulatory approaches will be required to achieve the necessary load reductions to meet the TMDL. Pennsylvania's program is being constructed to integrate State requirements (under Act 167) for stormwater management planning, Federal requirements for permitting through the NPDES program, and voluntary financial incentives provided to communities and project sponsors. Pennsylvania also recently adopted a Comprehensive Stormwater Management Policy (September 28, 2002).

Stormwater management was identified as a priority in Pennsylvania during 15 water forums held throughout the State during 2001. As a result, PA DEP proposed a Comprehensive Stormwater Management Policy (September 28, 2002) to more fully integrate post-construction stormwater planning requirements, emphasizing the use of groundwater infiltration and volume and flow rate control BMPs into the NPDES permitting program. The policy also emphasizes the obligation under Pennsylvania's water quality standards (25 Pa. Code Section 93.4a) for stormwater management programs to maintain and protect existing uses and the level of water quality necessary to protect those uses.

Under the NPDES storm water program, operators of large, medium and regulated small municipal separate storm sewer systems (MS4s) require authorization to discharge pollutants under an NPDES permit. The NPDES permitting program is implemented by the PA DEP under a delegation agreement with EPA.

Phase I of the Federal Stormwater NPDES Program began in 1990 and covered municipalities having a municipal separate storm sewer system and having a population greater than 100,000 (including portions of Philadelphia). Phase I also extended to construction activities which disturbed more than 5 acres of land and to 11 categories of industrial activity. In Pennsylvania, the City of Philadelphia is one of two cities covered under the Phase I program.

Phase II implementation is underway. Phase II requirements for the Federal NPDES stormwater program were described in Federal regulations at 40 CFR 122(a)(16) issued in December 1999. Phase II extended the requirement to small MS4s in urbanized areas as defined by the 1990 and 2000 census data and for construction activities requiring stormwater permits for one to five acres of disturbed land. As a result, the 16 municipalities in the Wissahickon Creek watershed are now being required to apply for and comply with NPDES permits for stormwater.

Implementation of the BMPs consistent with the stormwater management program and the "Minimum Control Measures" outlined in 40 CFR 132.34 is considered necessary to constitute compliance with the standard of compliance to the maximum extent practicable. To achieve reductions in stormwater discharges, EPA regulations establish six categories of BMPs that must be met by permittees (these are "narrative" permit effluent limitations). The six BMP categories, also called "minimum control measures" in the Federal regulations, are:

1.   Public education and outreach on stormwater impacts;
2.   Public involvement/participation consistent with state/local requirements in the development of a stormwater management plan;
3.   Illicit discharge detection and elimination, including mapping of the existing stormwater sewer system (including at least the outfalls) and adoption of an ordinance to prohibit illicit connections and control erosion and sedimentation from development;
4.   Control of runoff from construction sites when one to five acres of land are disturbed (Phase I of the Federal Stormwater NPDES Program covered sites larger than five acres);
5.   Post-construction stormwater monitoring and management in new development and redevelopment; and
6.   Pollution prevention and good housekeeping for municipal operations and maintenance facilities.

Under Phase II, permittees are also required to establish measurable goals for each BMP. Pennsylvania has also developed a protocol which MS4s covered under the general permit can adopt to satisfy the requirements of the permit. MS4s can also choose to develop their own programs, but they must seek PA DEP approval.

In order to carry out the Phase II NPDES Stormwater program, PA DEP developed a General Permit for Stormwater Discharges from Small MS4s (PAG-13) to provide NPDES coverage to the more than 700 municipalities in Pennsylvania, which EPA reviewed and approved. As described by PAG-13, the MS4 permittee must, within the permit term, implement and enforce a stormwater management program approved by PA DEP which is designed to reduce the discharge of pollutants from its MS4 to the maximum extent practicable, with the goal of protecting water quality and satisfying the appropriate water quality requirements of the Federal Clean Water Act and the Pennsylvania Clean Streams Law. The program must contain a schedule, BMPs, and measurable goals for the six Minimum Control Measures as described in the Federal regulations and in PAG-13.

In accordance with Phase II NPDES Stormwater requirements, the municipalities in the Wissahickon Creek watershed were required to apply for a permit by March 10, 2003 and are required to implement a stormwater management program by March 10, 2008. All have done so and their Notices of Intent are under review.

The first step to effectively address the complex and varied nature of this watershed is to develop a Watershed Management Plan which contains a plan of action for flow and pollutant load reduction and groundwater recharge. The Plan should address three major facets of watershed rehabilitation including: (1) flow and pollutant reduction mechanisms (structural and nonstructural BMPs), (2) institutional mechanisms (Memorandum of Agreements between municipalities and revised municipal ordinances); and (3) funding mechanisms (state and Federal grants, local utility fees etc.)

7.0   Public Participation

Public participation is not only a requirement of the TMDL process, but is essential to its success. At a minimum, the public must be allowed at least 30 days to review and comment prior to establishing a TMDL. Also, EPA must provide a summary of all public comments and responses to those comments to indicate how the comments were considered in the final decision.

Multiple publicly held meetings have been provided throughout all stages of the project to inform and update the public on all aspects of the project as it evolved. The public was encouraged to participate in data collection efforts and provide comments to a report of the data review and proposed TMDL methodology prior to TMDL development. In addition, EPA provided the public the unique opportunity to suggest modeling scenarios prior to TMDL development. As a result, several suggestions of stakeholders were included in TMDL development.

EPA also met with major dischargers on several occasions throughout and after the public comment period of the first draft Nutrient and Siltation TMDL Development for Wissahickon Creek, Pennsylvania to discuss options for nutrient TMDLs. These meetings provided stakeholders' opportunity to question EPA's contractor during technical review of the models and provided EPA with insight regarding model scenarios that could be tested for development of WLAs.

Following public comment, the draft Modeling Report for Wissahickon Creek, Pennsylvania Nutrient TMDL Development, the low-flow model utilized for development of nutrient TMDLs was revised to address concerns of stakeholders. Likewise, specific issues were addressed regarding calculation of siltation TMDLs. Due to the extent of modifications to the analytical framework resulting in subsequent changes in TMDL results and WLAs, the TMDL Report was re-opened for public comment on June 9th, 2003.

8.0   References

1. Ambrose, R.B., T.A. Wool, J.L. Martin, J.P. Connolly, and R.W. Schanz. 1991. WASP5.x, A Hydrodynamic and Water Quality Model - Model Theory, User's Manual, and Programmer's Guide. USEPA, Athens, GA.

2. Binger, R. L., F.D. Theurer, R.G. Cronshey, R.W. Darden. 2001. AGNPS 2001 Available from http://www.sedlab.olemiss.edu/AGNPS.html

3. Bingner, R. L., Murphree, C. E., Mutcher, C. K. 1989. Comparison of Sediment Yield Models on Watershed in Mississippi. Trans. ASAE 32, 529-544.

4. Bingner, R. L. 1990. Comparison of the Components Used in Several Sediment Yield Models. Trans. of the ASAE 33(4): 1229-1238.

5. Boyer, M.R. 1989. Aquatic Biological Investigation, Wissahickon Creek, 3/15/89 (unpublished internal report). PA DEP, Conshohocken, PA.

6. Boyer, M.R. 1997. Aquatic Biology Investigation, Wissahickon Creek, 6/17/97 (unpublished internal report). PA DEP, Conshohocken, PA.

7. Chapra, S.C. 1997. Surface Water Quality Modeling. The McGraw-Hill Company, INC., New York, USA.

8. Evans, B.M., S.A., Sheeder, K.J. Corrandi, and W.S. Brown. 2001. AVGWLF Version 4.0: User's Guide.

9. Everett, A.C. 2002. Periphyton Standing Crop and Diatom Assemblages in the Wissahickon Watershed. (unpublished internal report). PA DEP, Conshohocken, PA.

10. Haith, D.A., and L.L. Shoemaker. 1987. Generalized Watershed Loading Functions for Streamflow Nutrients. Water Resources Bulletin 23(3):471-478.

11. Haith, D.A., R. Mandel, and R.S. Wu. 1992. GWLF: Generalized Watershed Loading Functions User's Manual, Version 2.0. Department of Agriculture and Biological Engineering, Cornell University, Ithaca, NY.

12. Hamrick, J.M. 1992. A Three-Dimensional Environmental Fluid Dynamics Computer Code: Theoretical and Computational Aspects, Special Report 317. The College of William and Mary, Virginia Institute of Marine Science. 63 pp.

13. Maidment, D. R. (Editor in Chief). 1993. Handbook of Hydrology, McGraw-Hill, Inc.

14. PA DEP, 1997. Implementation Guidance for Determining Water Quality Based Point Source Effluent Limitations. Document ID: 391-2000-003.

15. PA DEP. 2002. Comprehensive Stormwater Management Policy.

16. Pennsylvania Code. Title 25, Environmental Protection. Web page: http://www.pacode.com/secure/data/025/025toc.html

17. Schubert, S.T. 1996. Aquatic Biology Investigation, Wissahickon Creek, 2/6/96 (unpublished internal report). PA DEP, Conshohocken, PA.

18. Shanaha, P. and M. Alam. 2001. The Water Quality Simulation Program, WASP5, Version 5.2-MDEP Manual:Part A. Hydraulics and Water Resource Engineers, INC., Waltham, MA.
19. Strekal, T.A. 1976. Aquatic Biology Investigation - Sewerage, Wissahickon Creek, 12/29/76 (unpublished internal report). PA DEP, Conshohocken, PA.
20. Tetra Tech, Inc. 2002. Data Review for Wissahickon Creek, Pennsylvania. Prepared for PA DEP and EPA Region 3.

21. Theurer, F.D. and R.G.Cronshey. 1998. AnnAGNPS—Reach Routing Processes. In Proceedings First Federal Interagency Hydrologic Modeling Conference, 19-23 April 1998, Las Vegas, NV.
22. USEPA. 2003. Nutrient and Siltation TMDL Development for Wissahickon Creek, Pennsylvania. U.S. Environmental Protection Agency, Region 3, Philadelphia, Pennsylvania.
23. Viessman, W. Jr. and G. L. Lewis. 1996. Introduction to Hydrology (4th Edition). HarperCollins College Publishers, New York, NY
24. Vogelmann, J.E., T. Sohl, and S.M. Howard. 1998. Regional characterization of land cover using multiple sources of data. Photogrammetric Engineering and Remote Sensing 64: 45-57.
25. West, N. 2000. Diatoms of Wisshickon Creek (PA), (unpublished internal report). Academy of Natural Sciences of Philadelphia, Philadelphia, PA.

Index